Fracture mechanics methodology

ENGINEERING APPLICATION OF FRACTURE MECHANICS
Editor-in-Chief: George C. Sih

Also in this series:

E.E. Gdoutos
Problems of mixed mode crack propagation
1984. ISBN 90 247 3055 4

Fracture mechanics methodology

Evaluation of Structural Components Integrity

Edited by

George C. Sih

Lehigh University,
Institute of Fracture and Solid Mechanics,
Packard Laboratory Building Number 19,
Bethlehem, PA 18015, USA

Luciano de Oliveira Faria

Technical University of Lisbon,
Lisbon, Portugal

1984 **MARTINUS NIJHOFF PUBLISHERS**
a member of the KLUWER ACADEMIC PUBLISHERS GROUP
THE HAGUE / BOSTON / LANCASTER

Distributors

for the United States and Canada: Kluwer Academic Publishers, 190 Old Derby Street, Hingham, MA 02043, USA
for the UK and Ireland: Kluwer Academic Publishers, MTP Press Limited, Falcon House, Queen Square, Lancaster LA1 1RN, England
for all other countries: Kluwer Academic Publishers Group, Distribution Center, P.O. Box 322, 3300 AH Dordrecht, The Netherlands

Library of Congress Cataloging in Publication Data

Main entry under title:

Fracture mechanics methodology.

   (Engineering application of fracture mechanics ;
v. 1)
   Includes bibliographical references and index.
   1. Fracture mechanics. 2. Airplanes--Materials.
I. Sih, G. C. (George C.) II. Oliveira Faria,
Luciano de. III. Series.
TA409.F7175 1984   620.1'126   83-27452
ISBN 90-247-2941-6

ISBN 90 247 2941 6 (this volume)
ISBN 90 247 3056 2 (series)

Copyright

PRINTED IN THE NETHERLANDS

Contents

Series on engineering application of fracture mechanics   VII

Foreword   IX

Editors' preface   XI

Contributing authors   XIII

Group photograph   XIV

Chapter 1.   Fatigue life prediction: metals and composites   i
      R. Badaliance

   1.1.   Introduction   1
   1.2.   Random spectrum load generation   2
   1.3.   Constant amplitude fatigue   3
   1.4.   Spectrum fatigue   9
   References   12

Chapter 2.   Fracture mechanics of engineering structural components   35
      G.C. Sih

   2.1.   Introduction   35
   2.2.   Strength and fracture properties of materials   36
   2.3.   Simple fracture experiments   44
   2.4.   Design of machine and structural components   50
   2.5.   Ductile fracture   61
   2.6.   Fatigue crack propagation   70
   2.7.   Appendix I. Strain energy density factor in linear elasticity   86
   2.8.   Appendix II. Critical ligament length   87
   2.9.   Appendix III. Fracture toughness test   88
   2.10. Appendix IV. A brief account of ductile fracture criteria   94
   References   99

Contents

Chapter 3.   Failure mechanics: damage evaluation of structural components   103
      O. Orringer

   3.1.   Introduction   103
   3.2.   Failure of a railroad passenger car wheel   104
   3.3.   Describing the load environment   106
   3.4.   Interpreting service load data   117
   3.5.   Predicting safe life   131
   3.6.   Maintaining perspective   143
   3.7.   Concluding remarks   144
   References   148

Chapter 4.   Critical analysis of flaw acceptance methods   151
      C.M. Branco

   4.1.   Introduction   151
   4.2.   Defects: distribution and non-destructive testing capability   151
   4.3.   Damage tolerance assessment   155
   4.4.   Flaw acceptance criteria   158
   4.5.   Conclusions   167
   References   167

Chapter 5.   Reliability in probabilistic design   169
      L. Faria

   5.1.   Introduction   169
   5.2.   Structural integrity   170
   5.3.   Designing for structural integrity   172
   5.4.   Safety factor and reliability   172
   References   174

Subject index   175

Series on engineering application of fracture mechanics

Fracture mechanics technology has received considerable attention in recent years and has advanced to the stage where it can be employed in engineering design to prevent against the brittle fracture of high-strength materials and highly constrained structures. While research continued in an attempt to extend the basic concept to the lower strength and higher toughness materials, the technology advanced rapidly to establish material specifications, design rules, quality control and inspection standards, code requirements, and regulations for safe operation. Among these are the fracture toughness testing procedures of the American Society of Testing Materials (ASTM), the American Society of Mechanical Engineers (ASME) Boiler and Pressure Vessel Codes for the design of nuclear reactor components, etc. Step-by-step fracture detection and prevention procedures are also being developed by the industry, government and university to guide and regulate the design of engineering products. This involves the interaction of individuals from the different sectors of the society that often presents a problem in communication. The transfer of new research findings to the users is now becoming a slow, tedious and costly process.

One of the practical objectives of this series on *Engineering Application of Fracture Mechanics* is to provide a vehicle for presenting the experience of real situations by those who have been involved in applying the basic knowledge of fracture mechanics in practice. It is time that the subject should be presented in a systematic way to the practicing engineers as well as to the students in universities at least to all those who are likely to bear a responsibility for safe and economic design. Even though the current theory of linear elastic fracture mechanics (LEFM) is limited to brittle fracture behavior, it has already provided a remarkable improvement over the conventional methods not accounting for initial defects that are inevitably present in all materials and structures. The potential of the fracture mechanics technology, however, has not been fully recognized. There remains much to be done in constructing a quantitative theory of material damage that can reliably translate small specimen data to the design of large size structural components. The work of the physical metallurgists and the fracture mechanicians should also be brought together by reconciling the details of the material microstructure with the assumed continua of the computational methods. It is with the aim of developing a wider appreciation of the fracture mechanics technology applied to the design of engineering structures such as aircrafts, ships, bridges, pavements, pressure vessels, off-shore structures, pipelines, etc. that this series is being developed.

Undoubtedly, the successful application of any technology must rely on the soundness of the underlying basic concepts and mathematical models and how they reconcile with each other. This goal has been accomplished to a large extent by the book series on *Mechanics of Fracture* started in 1972. The seven published volumes offer a wealth of information on the effects of defects or cracks in cylindrical bars, thin and thick plates, shells, composites and solids in three dimensions. Both static and dynamic loads are considered. Each volume contains an introductory chapter that illustrates how the strain energy criterion can be used to analyze the combined influence of defect size, component geometry and size, loading, material properties, etc. The criterion is particularly effective for treating mixed mode fracture where the crack propagates in a non-self similar fashion. One of the major difficulties that continuously perplex the practitioners in fracture mechanics is the selection of an appropriate fracture criterion without which no reliable prediction of failure could be made. This requires much discernment, judgement and experience. General conclusion based on the agreement of theory and experiment for a limited number of physical phenomena should be avoided.

Looking into the future the rapid advancement of modern technology will require more sophisticated concepts in design. The micro-chips used widely in electronics and advanced composites developed for aerospace applications are just some of the more well-known examples. The more efficient use of materials in previously unexperienced environments is no doubt needed. Fracture mechanics should be extended beyond the range of LEFM. To be better understood is the entire process of material damage that includes crack initiation, slow growth and eventual termination by fast crack propagation. Material behavior characterized from the uniaxial tensile tests must be related to more complicated stress states. These difficulties could be overcome by unifying metallurgical and fracture mechanics studies, particularly in assessing the results with consistency.

This series is therefore offered to emphasize the applications of fracture mechanics technology that could be employed to assure the safe behavior of engineering products and structures. Unexpected failures may or may not be critical in themselves but they can often be annoying, time-wasting and discrediting of the technical community.

Bethlehem, Pennsylvania   G.C. Sih
1984   Editor-in-Chief

Foreword

This book consists of a collection of lectures prepared for a short course on "Fracture Mechanics Methodology" sponsored by the Advisory Group for Aerospace Research and Development (AGARD), part of the North Atlantic Treaty Organization (NATO). The course was organized jointly by Professor George C. Sih of the Institute of Fracture and Solid Mechanics at Lehigh University in the United States and Professor Luciano Faria from Centro de Mecanica e de Materiais das Universidade de Lisboa in Portugal. It was held in Lisbon from June 1 to 4, 1981. Dr. Robert Badaliance from the McDonnell Aircraft Company in St. Louis and Dr. Oscar Orringer from the Department of Transportation in Cambridge are the other US lecturers while Professor Carlos Moura Branco from Portugal also lectured. The audience consisted of engineers from the Portuguese industry with a large portion from the aeronautical sector and others who are particularly interested to apply the fracture mechanics discipline for analyzing the integrity of structural components and fracture control methods. Particular emphases were given to the fundamentals of fracture mechanics as applied to aircraft structures.

The Portuguese Fracture Group headed by Professor L. Faria should be acknowledged for making the local arrangements. Mr. C.E. Borgeaud, Chief of Plans and Programmes and Colonel J.C. de Chassey (FAF), Deputy for Plans and Programmes, both of whom are from the NATO Headquarters in France, coordinated the paper work. To this end, Colonel R. Grossel from the US Air Force in The Pentagon also contributed. Finally, the publication of this book was made possible to the public only upon the approval of NATO.

Lisbon, Portugal   G.C. Sih
June, 1981   Coordinator

Editors' preface

Although fracture mechanics has been in use for more than two decades to characterize the resistance of materials against crack growth, its potential application to structure design has still not been fully explored. The concept of durability and damage tolerance adopted in aircraft structure design could not be implemented without a knowledge of fatigue crack growth under service loading conditions. Frequently, insufficient attention is given to the translation of data collected from laboratory specimens to the design of structural components. This can be attributed to inadequacy of stress and/or failure analysis. The importance of applying fracture mechanics for evaluating structure integrity cannot be overemphasized. It is with this objective in mind that the present book is offered to the technical community.

Chapter 1 is concerned with the application of fracture mechanics for analyzing aircraft structural components made of metallic and composite materials. Based on the strain energy density theory, the rate of crack propagation is assumed to be a function of the strain energy density factor range during fatigue spectrum loading. The effects of stress amplitude and mean stress are accounted for explicitly. Characterized and correlated are constant amplitude fatigue crack growth data of several aluminum alloys (7049-T73, 7075-T6, 7075-T73, 2219-T851), titaniums (Ti-6Al-4V mill and beta annealed), and steels (HP9-4-.30 and 300M). These results are then used in conjunction with a cycle-by-cycle integration scheme to predict the spectrum fatigue lives of panels with center cracks and surface flaws subjected to a fighter aircraft landing gear load history. Mixed mode fatigue crack growth from a notch is also discussed with emphases placed on predicting crack trajectories from the strain energy density criterion.

The remaining portion of Chapter 1 deals with the durability and life prediction of composite materials used in aircraft structures. Examined in particular is their behavior under a variety of loading conditions found in a fleet of multi-mission military aircraft. The damage pattern in a composite can no longer be represented by a single dominant flaw configuration as in the case of a metal. A correlation parameter based on the strain energy density theory modeling microcracks in the laminate matrix is developed. Predictions on the life of several laminates subjected to various load spectra are made from the constant amplitude fatigue data and the results are compared with the corresponding experimental data.

Emphases on predicting different types of fracture failure modes by application of the strain energy density theory* are presented in Chapter 2. Under service conditions where the orientation between the load direction and crack plane may constantly change, crack growth is likely to occur along a curve path in a non-self-similar fashion. Unless these growth characteristics are accounted for in the analysis, no confidence can be placed in predicting the life of aircraft structural components. It is shown how data collected from Mode I fracture toughness testing can be used to forecast the allowable load for the more complex structural components. The phenomenon of slow crack growth is also analyzed by application of the strain energy density criterion. This process is usually accompanied by plastic deformation and is obviously load history dependent. In other words, the crack growth behavior depends sensitively on the applied load increments. An example of a three-dimensional through crack growing slowly in an incrementally loaded plate is provided. Quantitative results for three different plate thicknesses are reported illustrating the interaction between yielding and crack growth. The critical strain energy density factor is shown to remain constant even when the plate thickness and plastic deformation are varied.

Fatigue crack growth is discussed in context with the accumulation of strain energy density in an element ahead of the crack. A threshold value is assumed to exist for the hysteresis strain energy density function which represents the energy accumulated for one increment of crack growth. Such an approach can be applied to each point along a crack border of arbitrary shape under general loading conditions. Obviously, the analysis must involve the theoretical modeling of a material that dissipates and/or accumulates energy during each cycle of loading. The continuum theory of plasticity has been commonly used for modeling the dissipative behavior of metal alloys. Simplified and/or linearized fatigue crack growth relations can only be developed on the basis of an in-depth understanding of the nonlinear and path dependent process of material failure by fatigue.

The title of Chapter 3 contains the terminology, "Failure Mechanics", which is adopted to include damage modes other than sharp cracks. Crack growth may not always be the predominant damage mode in structures. Limit load is one such example. This Chapter focuses attention on actual cases of structural failures involving fleets in service. Shortcuts and approximate methods are necessary for arriving at a resolution and defining course of actions. The failure of a railroad passenger car wheel is used as an example. The analysis involves a knowledge of the load distribution and sequence as they can significantly influence fatigue crack nucleation life and subsequent crack growth. Spectral analysis is essential as fatigue life is load history dependent. Requirements for the fighter aircrafts obviously differ from those for the transport aircrafts. The interpretation of service load data is covered in detail. Prediction on the safe life of structural components is also discussed in connection with a linear damage rule and the conventional fatigue crack growth relation based on the Mode I stress intensity factor range.

Chapter 4 considers the quality control aspects of materials as it plays an important role in structure design. Since mechanical imperfections or flaws are inherent in all

---

*Unlike the conventional fracture mechanics approach based on energy release rate or critical stress intensity factor that is restricted to crack behavior in the linear elastic range, the strain energy density approach applies to all classes of engineering materials and loading conditions such that the crack can grow non-self-similarly in the presence of plastic deformation.

man-made materials, their distribution and size should be controlled or known during the initial design stage. The "Flaw Acceptance Methodology" is discussed in this context. Implementation of the method requires the non-destructive detection of the position and size of flaws so that their effects on the structure integrity can be evaluated. Life prediction on the remaining life of any structural component depends, of course, on the appropriate criterion chosen in the analysis. Discussed in this Chapter is a tolerable defect parameter related to the crack tip opening distance. Emphases are given to applying the flaw acceptance method to aircraft structures.

The fifth and final Chapter in this book deals with the probabilistic approach for addressing reliability in structure design. The factors to be considered for maintaining aircraft component reliability involve effective inspection procedure, type of materials used, static and dynamic load safety factors, locations of stress concentrations and load transfer characteristics. Fracture mechanics considerations are also given in terms of fatigue crack growth and damage tolerance. The necessity for establishing confidence levels in translating specimen test data to the design of allowable loads on structures is stressed.

The five Chapters in this book provide not only a basic knowledge of the fracture mechanics discipline but with application to design involving real structures in service. The examples stress the importance of material defects from which failure initiates. The editors wish to thank the contributors of this volume for completing the manuscript.

Lisbon, Portugal
June, 1981

C.G. Sih
L. Faria
Volume Editors

Contributing authors

R. Badaliance
Structural Research Department
McDonnell Aircraft Company
St. Louis, Missouri, USA

C.M. Branco
Department of Mechanics of Materials
University of Minho
Largo do Paco, 4719 Braga Codex
Portugal

L. Faria
Center of Mechanics and Materials
University of Lisbon
Lisbon, Portugal

O. Orringer
Mechanical Engineer Staff
US Department of Transportation
Transportation Systems Center
Cambridge, Massachusetts, USA

G.C. Sih
Institute of Fracture and Solid Mechanics
Lehigh University
Bethlehem, Pennsylvania, USA

Participants of a short course on Fracture Mechanics Methodology sponsored by AGARD, June 1981, Lisbon, Portugal.

Fatigue life prediction: metals and composites

1.1. Introduction

The structural design philosophy based on durability and damage tolerance requires prediction of fatigue crack growth due to anticipated service loads. The essential elements of this analysis are: service load histories (random spectrum loads); characterization of constant amplitude data for material in question; and fatigue model. The development and application of this analysis to aircraft structures made of metallic and composite materials are discussed.

The strain energy density factor is used to characterize and correlate the fatigue crack growth behavior of metallic materials. It is postulated that the rate of crack propagation is a function of the range of this factor during fatigue spectrum loading. This characterization provides the means to account for stress range and mean stress levels explicitly. Furthermore, it enables one to use uniaxial (Mode I) test data to predict mixed mode fatigue crack growth. The strain energy density factor range was used to correlate Mode I constant amplitude subcritical crack growth data of certain aluminums (7049-T73, 7075-T6, 7075-T73, 2219-T851), titaniums (Ti-6Al-4V mill and beta annealed), and steels (HP9-4-.30 and 300M). These constant amplitude curves are then used in conjunction with cycle-by-cycle integration schemes to predict spectrum fatigue lives of center cracks and semi-elliptic surface flaws in plates subjected to a fighter aircraft landing gear stress history. In addition, experimental results for mixed mode fatigue crack propagation in a plate containing a center crack are discussed. The strain energy density factor range was used to predict fatigue life and crack growth path for sharp cracks and cracks emanating from a notch under mixed mode conditions.

The introduction of composite materials into aircraft structures has raised the question of durability and life prediction under the variety of loading conditions found in a fleet of multimission military aircraft. The behavior of composite laminates under fatigue loading is quite different from that of metallic materials. One difference is that in composite materials there is no single dominant flaw which grows and leads to eventual failure. The approach taken in quantifying fatigue life degradation in composites is based on the assumption that the matrix is the weak link of the system. A correlation parameter based on the concept of strain energy density factor for micro-cracks in the laminate matrix is developed. This correlation parameter and constant amplitude fatigue data were used in conjunction with linear residual strength reduction fatigue model to predict the life of different laminates subjected to various spectra. These results are then compared with the corresponding experimental data.

1.2. Random spectrum load generator

In the design of an aircraft to meet a predefined life requirement, a realistic knowledge of anticipated loading conditions to which the structure would be subjected is essential. This information is necessary for fatigue analysis and representative testing of structural components and of the full scale airframe.

A review of fighter aircraft load usage time histories [1–6] indicates much more of a random sequence than an orderly application. This has led to random load fatigue design and testing in the aircraft industry. Fatigue tests [7–9] to determine effects of load sequencing (Lo-Hi, Hi-Lo), block size, and negative loads have shown that lives with ordered loads are substantially different from random load test data. Results of fatigue tests conducted at McDonnell Aircraft Company to determine effects of load sequencing are summarized in Tables 1.1 and 1.2. These data indicate that random load sequences need to be used to insure that the fatigue results are representative of actual service usage.

Table 1.1.   Spectrum fatigue life variations in aluminum specimens with open holes.

	Spectrum definition	Test life laboratory spectrum hours
Spectrum A	Random sequence without unloading	4300
	Random sequence with unloading	3040
	Random sequence with negative loads	1750
	Lo-Hi sequence with unloading	2700
	Lo-Hi sequence with unloading truncated at 85%	1510
Spectrum B	Lo-Hi sequence with 1 g minimum	1420
	Lo-Hi sequence with − 1 g minimum	680
	Lo-Hi sequence with − 3 g minimum	380

The work done in conjunction with the development of aircraft maneuver loading [10–13] has shown that random noise theory results correlate well with measured flight load factor histories. These were developed based on the mathematical analysis of random noise developed by S.O. Rice [14]. It was pointed out in [15] that a common fault of other methods of load factor history simulation is the arbitrary and unrealistic coupling of peaks and valleys. An advantage of the random noise theory approach is that both the exceedance content and the frequency content of the process can be preserved. The preservation of the frequency content enables the proper coupling of peaks and valleys. The details of applying the random noise theory to development of random load histories are given in [4]. Once the proper random stress history is developed, it can be used in conjunction with properly characterized constant amplitude fatigue to predict spectrum fatigue life of structural components.

Table 1.2.   Spectrum fatigue life variations in titanium specimens with open holes

	Spectrum definition	Test life laboratory spectrum hours
	Random sequence of peaks 250 hr block size. Highest load = 131.6%	94750
Spectrum C	Random sequence of peaks 250 hr block size. Highest load = 111.8%	55250
	Lo-Hi sequence of peaks 250 hr block size. Highest load = 131.6%	24250

During the developmental stages of a fighter structure (design and testing) the basic parameters involved in the development of the baseline spectrum such as mission mixes (air-to-air, air-to-ground) can be varied and the influence of these variations on life determined. In Section III the influence of a number of these variations on spectrum fatigue life are discussed in detail.

1.3. Constant amplitude fatigue

In this section, the constant amplitude fatigue crack growth life of metallic materials and total fatigue life of graphite/epoxy composite laminates are discussed.

Fatigue crack growth in metals

Currently accepted fatigue crack growth laws are based on the premise of relationship between crack growth rate (da/dN) and the stress intensity factory range (ΔK). Although this type of correlation accounts for the stress range in a fatigue load cycle, it fails to account explicitly for mean stress, an important parameter. Furthermore, they are limited to a uniaxial stress field and a single mode of crack propagation. These problems are overcome by considering the change in strain energy density at the crack tip [16, 17, 18]. In this work, it is postulated that the rate of the crack propagation da/dN is a function of the strain energy density factor range ΔS. In a functional form, this can be written as:

$$\frac{da}{dN} = f(\Delta S) \tag{1.1}$$

with da/dN approaching infinity at the onset of instability. The strain energy density factor range is defined as:

$$\Delta S = S_{max} - S_{min} \tag{1.2}$$

where S_{max} and S_{min} are the strain energy density factors associated with maximum and minimum stresses. The strain energy density factor, S, is given in References [19] and [20] as:

$$S = a_{11}k_1^2 + 2a_{12}k_1k_2 + a_{22}k_2^2 \tag{1.3}$$

3

where k_1 and k_2 are the Mode I and Mode II stress intensity factors.* The coefficients a_{ij} contain the elastic constants (μ, κ) and vary with the polar angle, θ, measured from the crack tip. These coefficients are:

$$a_{11} = \frac{1}{16\mu} \{(1 + \cos\theta)(\kappa - \cos\theta)\}$$

$$a_{12} = \frac{1}{16\mu} \{\sin\theta\,(2\cos\theta - (\kappa - 1))\} \qquad (1.4)$$

$$a_{22} = \frac{1}{16\mu} \{(\kappa + 1)(1 - \cos\theta) + (1 + \cos\theta)(3\cos\theta - 1)\}.$$

The elastic constant, κ, is equal to $(3 - 4\nu)$ for plane strain and $(3 - \nu)/(1 + \nu)$ for generalized plane stress, while μ stands for the shear modulus of elasticity.

Consequently, the range of the strain energy density factor for a given load cycle can be written as:

$$\Delta S = a_{11}(k_{1\max}^2 - k_{1\min}^2) + 2a_{12}(k_{1\max}k_{2\max} - k_{1\min}k_{2\min})$$
$$+ a_{22}(k_{2\max}^2 - k_{2\min}^2). \qquad (1.5)$$

For Mode I fatigue crack growth, the strain energy density factor range can be written as

$$\Delta S = a_{11}(1 - R^2)k_{1\max}^2 \qquad (1.6)$$

where $a_{11} = (\kappa - 1)/8\mu$, $R =$ stress ratio $(\sigma_{\min}/\sigma_{\max})$ and $k_{1\max}$ is the maximum stress intensity factor.

This strain energy density factor range is used to correlate the constant amplitude fatigue crack growth data for several different materials [16]. Correlation of constant amplitude crack growth rates for mill-annealed titanium (Ti-6Al-4•/) with ΔS and ΔK for three different stress ratios $(R = .02, .5, -1)$ are shown in Figures 1.1A and 1.1B. Correlation of crack growth rates of Beta-annealed titanium with ΔS is given in Figure 1.2. The correlations with ΔS show negligible dependence on stress ratio, R, while the correlation with ΔK shows strong dependence on R. In these correlations negative stress intensity factors were truncated to zero.

However, for steels 300M (Figure 1.3A) and HP9-4-.30 (Figure 1.4A), and aluminums 7075-T73 (Figure 1.5A), 7075-T6, 7049-T73 and 2219-T851, correlations with ΔS do indeed demonstrate dependence on stress ratio.

An empirical parameter, α, was devised to account for this dependency on stress ratio. This parameter is given as:

$$\alpha = (1 + R)\bigg/\left\{1 + R\cdot\left[\frac{\sigma_u(1 + \%RA)}{\sigma_y}\right]^2\right\} \qquad (1.7)$$

where $\%RA =$ percent reduction in area, σ_u is the ultimate stress and σ_y is the yield stress of a standard tensile specimen, and R is the stress ratio. With this factor included for different stress ratios, the experimental data collapses onto a single curve for each material. Material properties necessary for this correlation are given in Table 1.3, for the materials considered here. The crack growth rates of steels, 300M, HP9-4-.30 and

---

*The factor k, differs from K, whose critical value is referred as the fracture toughness by a factor $\sqrt{\pi}$, i.e., $K_1 = \sqrt{\pi}k_1$.

aluminum 7075-T73 as a function of $\alpha\Delta S$ are given in Figures 1.3B, 1.4B and 1.5B, while crack growth rates of aluminums 7075-T6, 7049-T73 and 2219-T851 as a function of $\alpha\Delta S$ are given in Figures 1.6, 1.7 and 1.8.

These correlations show that the variation of crack growth rate da/dN with $\alpha\Delta S$ is of sigmoidal form. As a functional form it can be written as $da/dN = f(\alpha\Delta S)$. A constrained regression technique was used to develop the following relationship for da/dN as a function of $\alpha\Delta S$.

$$\frac{da}{dN} = \exp\left\{ C_1 \frac{[A \ln(\alpha\Delta S) - B]}{|A \ln(\alpha\Delta S) - B|} |A \ln(\alpha\Delta S) - B|^n \right.$$

$$\left. + C_2 [A \ln(\alpha\Delta S) - B]^2 + C_3 [A \ln(\alpha\Delta S) - B] + C_4 \right\}. \qquad (1.8)$$

The constants A, B, C_1, C_2, C_3, C_4 and the exponent n obtained from a best fit to the experimental data are given in Table 1.4. The form of the above equation does not represent a particular crack growth law, it merely provides a convenient mean in correlating the data. The important point is that with this concept the amount of testing can be minimized, and constant amplitude Mode I data can be used to predict mixed mode and spectrum crack propagation.

Table 1.3. Material properties

Material	E(10^6_{psi})	ν	σ_{yield} (ksi)	σ_{ult} (ksi)	% Reduction in area	β
Titanium (Ti-6Al-4V) mill-annealed	18.6	0.31	133	137	22	1.256
Titanium (Ti-6Al-4V) beta-annealed	17.8	0.31	125	135	14	1.231
Aluminum 7049-T73	10.1	0.33	72	81	37	1.541
Aluminum 7075-T6	10.3	0.33	78	85	33	1.450
Aluminum 7075-T73	10.2	0.33	67.5	76	30	1.464
Aluminum 2219-T851	10.5	0.33	51	66	19	1.540
Steel HP9-4-0.30	28.3	0.32	202	229	60	1.874
Steel 300M	29.3	0.30	235	293	46	1.759

$\beta = \sigma_u(1 + \%RA)/\sigma_y$

Mixed mode crack growth

To demonstrate that the above concept can be used to predict mixed mode fatigue crack growth in metallic materials, a number of simple mixed mode fatigue tests were performed [18] on center cracked plates, where the crack was at an angle with the load direction. The test specimens configuration is shown in Figure 1.9, while the test conditions are outlined in Table 1.5. These specimens were subjected to constant amplitude sinusoidal fatigue loads. A photograph of the 300M steel specimen after failure is shown in Figure 1.11, where it clearly shows that the fatigue crack growth originated at a location other than the apex of the elox notch. A photograph of the aluminum specimen tested at a stress ratio of $R = -\infty$ (compression only) is shown in Figure 1.12. The fatigue cracks grew towards the center of the specimen. This specimen did not fail in fatigue; in an attempt to expose the crack during the application of tensile load, the specimen failed at the elox slot.

Table 1.4   Table of numerically evaluuationed constants for crack growth rate

$$\frac{da}{dN} = \exp\left\{ C_1 \frac{[A \ln (\alpha \Delta S) - B]}{|A \ln (\alpha \Delta S) - B|} \, |A \ln (\alpha \Delta S) - B|^n \right.$$

$$\left. + C_2 [A \ln (\alpha \Delta S) - B]^2 + C_3 [A \ln (\alpha \Delta S) - B] + C_4 \right\}.$$

Material	n	A	B	C_1	C_2	C_3	C_4
Titanium (Ti-6Al-4V) mill annealed	2.700	0.240	0.980	12.2712	17.6460	13.8740	− 5.2722
Titanium (Ti-6Al-4V)* beta-annealed	2.250	0.250	0.871	35.0864	37.3560	11.4891	− 7.2694
Aluminum 7049-T73	2.090	0.300	0.770	8.8840	9.2742	6.4330	− 8.0720
Aluminum 705-T6	2.180	0.340	0.900	16.7503	20.2560	10.2530	− 6.2997
Aluminum 7075-T73	2.300	0.320	0.905	14.5210	19.2423	10.4151	− 7.0122
Aluminum 2219-T851†	2.600	0.400	0.460	4.0178	7.4676	9.1361	− 8.1061
Steel HP9-4-0.30	2.300	0.290	0.806	5.8938	7.1328	6.4986	− 9.1020
Steel 300M	2.140	0.360	0.894	18.5062	21.0322	6.3561	− 9.5792

*Compact tension specimen.
†Dehumidified Argon environment.

Table 1.5.   Summary of specimen and test condition.

Material	Specimen no.	Type of starter flaw	Flaw inclination angle (β)	Maximum fatigue stress (ksi)	Stress ratio (R)
7075-T7351	1	Elox-Slot	60°	15	0.02
7075-T7351	2	Elox-Slot	45°	15	0.02
7075-T7351	3	Elox-Slot	30°	15	0.02
7075-T7351	4	Elox-Slot	30°	15	− 1
7075-T7351	5	Elox-Slot	30°	15	− 1
7075-T7351	6	Elox-Slot	45°	15	− ∞
7075-T7351	7	Fatigue crack	45°	15	0.02
300M	8	Elox-Slot	60°	20	0.02

A cycle-by-cycle integration of fatigue crack growth rate, da/dN as a function of $\alpha \Delta S$, i.e.,

$$a_N = a_i + \sum_{n=1}^{n} \Delta a_n \qquad (1.9)$$

was used to calculate the mixed mode crack growth of the precracked aluminum specimen (Figure 1.10) and the 300M steel specimen with the elox slot (Figure 1.11). The value a_i is the initial crack length, and $\Delta a_n = da/dN$ for $\alpha \Delta S$ in cycle n of the stress history under consideration. In application of this methodology, the difficulty lies in the determination of stress intensity factors k_1 and k_2 for a curved branch crack which is continuously changing shape with the application of each load cycle. The determination of exact values of k_1 and k_2 is beyond the scope of this paper. The objective here is to demonstrate that uniaxial fatigue crack growth data can be used to predict mixed mode fatigue crack propagation. For the purpose of this demonstration,

Table 1.6.   Ultimate strength results.

Layup	Compression		Tension	
	Strain (μ in./in.)	Stress (ksi)	Strain (μ in./in.)	Stress (ksi)
48/48/4	8010	79.00	5638	61.87
25/67/8	9398	63.39	5848	42.91
16/80/4	10432	50.60	7293	38.08

the only singluar points considered are the growing crack tips; furthermore, the width effect of the specimen on the crack growth has been ignored by letting $k_{1n} = \sigma \sin^2 \beta_n \sqrt{a_n}$ and $k_{2n} = \sigma \sin \beta_n \cos \beta_n \sqrt{a_n}$ (Figure 1.13). These stress intensity factors in conjunction with the postulate of strain energy density criterion [19] determine the angle along which the initial crack growth will take place. Once the direction of the crack growth is determined, the strain energy density factor range corresponding to this angle is used to predict the amount of crack growth for a given stress ratio and number of cycles.

This analytic approach was used to predict the mixed mode constant amplitude fatigue crack growth of the precracked aluminum 7075-T7351 specimen (No. 7). The results of this prediction along with the experimental data are shown in Figure 1.14, indicating fairly good correlation considering the approximations of the stress intensity factors k_1 and k_2.

To predict fatigue crack growth of specimens with elox starter flaws requires slightly different consideration, since the elox flaw tip is more of a notch than a sharp crack, and the experimental evidence (Figures 1.11 and 1.12) indicate that the fatigue crack growth did not originate at the apex of the notch. The surface layer energy concept of [20] and [21] was used to predict the location around the notch where the crack would initiate, and the strain energy criterion was used to predict the direction at which the crack would grow. The predicted location and direction for crack initiation of the 300M steel specimen is shown in Figure 1.14. The radius of curvature at the end of the elox slot was measured to be approximately .125 inches, and was used to predict the initiation angle. The predicted and experimental data correlated quite favorably.

Fatigue of composite laminates

The fatigue damage mechanism in composite laminates is quite different from metallic structure. A fundamental difference is that the metallic structural component's fatigue life is governed by the initiation of a crack and its subsequent growth leading to instability and failure. Whereas fatigue life of composite laminates are not governed by a single dominant flaw, rather a multitude of cracks accumulate, leading to eventual failure of the laminate. In [22, 23] enhanced X-ray radiography was used to facilitate observation of this fatigue damage accumulation in graphite/epoxy laminate specimens with circular fastener holes. Tetrabromoethane or zink iodine-enhanced X-ray radiographs [23] show that matrix cracking is the dominant degradation mechanism over the greater part of the fatigue life of nondelamination prone graphite/epoxy laminates and structures. This damage progression sequence begins with matrix cracking (intralaminar) at the fiber-matrix interface within a ply followed by delamination in areas which have accumulated extensive cracking. These intralaminar matrix cracks and delamination interact to degrade the matrix, resulting in loss of fiber support leading to overall failure by microbuckling or rupture of the fibers. A recent literature survey [24] outlines research activities in the area of fatigue damage in composites.

These observations clearly indicate that the matrix is the weak link of the composite systems, with intralaminar cracking being the dominant damage mechanism. This damage mechanism can be modeled by [25] an isotropic strip of resin containing a through-the-thickness crack sandwiched between two semi-infinite orthotropic plates. This model is described in Figure 1.16, where E and ν are the elastic constants of the resin, while E_1, E_2, ν_{12}, and μ_{12} are the elastic constants of the equivalent orthotropic material surrounding the resin strip. The farfield stresses σ_1, σ_2, and σ_{12} are classical lamination plate theory stresses which assume a uniform strain distribution through-the-thickness of the laminate. The subscripts 1 and 2 represent directions parallel and perpendicular to the ply fibers, respectively. The strip width $2h$ is related to the fiber volume fraction. For low fiber volume fraction this width can be estimated to be $h = R(\frac{1}{2}\sqrt{(\pi/V_f)} - 1)$, where R is the fiber radius and V_f is the fiber volume fraction. However, for a graphite/epoxy composite system such as AS/3501-6 with high fiber volume fraction, the strip width $2h$ is very small, hence the limiting value $(h/a \rightarrow 0)$ can be used.

The intensity of the crack tip stress field for this problem can be written as:

$$k_1 = \Phi(1)\sigma_2\sqrt{a}$$
$$k_2 = \Psi(1)\sigma_{12}\sqrt{a}. \tag{1.10}$$

The functions $\Phi(1)$ and $\Psi(1)$ depend on the geometry and elastic properties of the composite, and are obtained numerically by solving the Fredholm integral equations given in [26]. The values of $\Phi(1)$ and $\Psi(1)$ for AS/3501-6 $(h/a \rightarrow 0)$ graphite/epoxy are:

$$\Phi(1) = .2843, \qquad \Psi(1) = .0911. \tag{1.11}$$

These stress intensity factors in conjunction with the strain energy factor S can be used to correlate the constant amplitude fatigue life data. The strain energy density factor S for two-dimensional problems is given as:

$$S = a_{11}k_1^2 + 2a_{12}k_1k_2 + a_{22}k_2^2 \tag{1.12}$$

where the coefficient a_{ij} are in terms of elastic constants and crack trajectory angle θ described in an earlier section.

In the development of correlation parameter, it was assumed that through-the-thickness cracks in the matrix grow parallel to adjacent fibers, i.e., $\theta_{cr} = 0$. This assumption was made for simplicity, the fact that the graphite/epoxy composites have a high fiber volume density which may force the matrix crack to propagate parallel to the fibers under cyclic loads. The strain energy density factors for each ply are summed as a damage indicating parameter, and normalized with respect to ultimate compressive or tensile strength of the composite laminate resulting in the following relationship:

$$\bar{S} = \left(\frac{\sigma_i}{F_{u_i}}\right)^2 \sum_{n=1}^{N} (S_{\text{per ply}})_i + \delta_{ij} \left(\frac{\sigma_j}{F_{u_j}}\right)^2 \sum_{n=1}^{N} (S_{\text{per ply}})_j \qquad (1.13)$$

where i, j = compression, tension

$$\delta_{ij} = \begin{cases} +1 & i \neq j \text{ (Tension–Compression)} \\ -1 & i = j \text{ (Compression–Compression)} \end{cases} \qquad (1.14)$$

where σ_i and σ_j are the farfield stresses applied to the laminate, while F_{u_i} and F_{u_j} are the ultimate strengths (compressive or tensile) of the laminate, with N being number of plies in the laminate.

The parameter $\bar{S}$ was used in [22] and [23] to correlate constant amplitude fatigue data for three different laminates made of AS/3501-6 graphite/epoxy. The three laminates considered are fiber dominated (48/48/4), intermediate (25/67/8) and matrix dominated (16/80/4) made of 10.4 Mil prepreg. The first number in the parenthesis is the percentage of $0°$ plies, the second number is the percentage of $\pm 45°$ plies and the third number is the percentage of $90°$ plies in the laminate. These layups are balanced and symmetric (for detail information concerning their stacking sequence see [23]). Constant amplitude fatigue results of these three laminates are shown in Figure 1.17. A simple single hole test specimen (Figure 1.19) was used to obtain these data. Specimens were tested with protruding-head close-tolerance bolts installed into the fastener hole, with the nut torqued finger-tight. Ultimate strength results of these three layups given in Table 1.6 were used in developing the correlation parameter $\bar{S}$.

The results shown in Figure 1.18 indicate that the parameter $\bar{S}$ correlates constant amplitude fatigue life for different layups and a given stress ratio. However, it falls short in correlating the stress ratio effect. An empirical relation $F(R)$ shown in Figure 1.20 was developed to account for stress ratio effect. The data used for developing $F(R)$ is the value of $\bar{S}$ for different stress ratios normalized with respect to the $\bar{S}$ of pure compression cycle (i.e., $R = -\infty$). With the aid of this empirical relationship, the constant amplitude fatigue data collapses onto a single line as shown in Figure 1.21 where $\bar{S}_{\text{eq}} = \bar{S}/F(R)$, which can be used to predict life for various layups and stress ratios.

1.4 Spectrum fatigue

In this section, the spectrum fatigue crack growth prediction for metals and spectrum fatigue life prediction for composite laminates are discussed.

Spectrum fatigue crack growth in metals

In predicting the crack growth life of metallic structural components due to arbitrary spectrum loads, it is essential to account for the stress range ($\Delta\sigma = \sigma_{\text{max}} - \sigma_{\text{min}}$) and mean stress level $\sigma_{\text{mean}} = 1/2(\sigma_{\text{max}} + \sigma_{\text{min}})$ in each load cycle. The concept of strain energy density factor range presented in the previous section explicitly accounts for these variables, which enables the analyst to use a simple cycle-by-cycle integration of crack growth to predict fatigue crack growth life.

To demonstrate the applicability and simplicity of this concept, fatigue crack growth was predicted for different materials with different types of cracks subjected to spectrum loads. A cycle-by-cycle integration of fatigue crack growth as a function of $\alpha\Delta S$ was used in conjunction with the following equation:

$$a_N = a_i + \sum_{n=1}^{N} \Delta a_n. \tag{1.15}$$

The value a_i is the initial crack length, and $\Delta a_n = \mathrm{d}a/\mathrm{d}N$ for $\alpha\Delta S$ in cycle n of the stress history under consideration.

This methodology was applied to predict spectrum fatigue crack growth in center cracked plates (Figure 1.22) of HP9-4-.30 and 300M steels and 7049-T73 aluminum, and a plate specimen of HP9-4-.30 steel with an elliptic surface flaw (Figure 1.23) subjected to in-plane loads. The stress intensity factor, k_1, for the center cracked plate was taken from [27], and solution for the elliptic surface flaw from [28]. The arbitrary load spectrum used for these calculations is a flight-by-flight stress history of the main landing gear of a fighter aircraft. Stress exceedance as a function of percent maximum stress is shown in Figure 1.24. Predicted crack growths of the center cracked plates are shown in Figures 1.25, 1.26 and 1.27. Crack growth retardation due to tensile over-loads and possible crack growth accelerations due to compressive loads were ignored in the predictions. As it can be seen, the correlation between the experimental data and analytic prediction is quite good.

The predicted crack growth of the elliptic surface flaw is shown in Figure 1.28, along with experimental data on two specimens. Again the prediction corresponds quite well with the test data. Photomacrographs of the fracture surfaces of the two elliptic surface flaw specimens are presented in Figure 1.29, showing the progression of the crack shape during the fatigue life.

Spectrum fatigue life of composites

As it was pointed out in a previous section, the fatigue damage process in composites is the accumulation of matrix cracks, which makes it amenable to the use of cumulative damage models for life prediction. In using these models, it is assumed that a predictable segment of the life is used with the application of every load cycle. The rate at which life is predicted to be degraded depends on the assumed damage indicating parameter and constant amplitude fatigue data. The constant amplitude fatigue predictive methodology developed in previous sections is used here in conjunction with two linear fatigue damage models to predict spectrum fatigue life. The fatigue damage models used are Miner's Rule and a linear strength reduction model [29].

The linear residual strength reduction model shown schematically in Figure 1.30, is a modified version of Miner's Rule with the assumption of a linear reduction in the residual compressive strength of the laminate due to cycle loading. Mathematically this can be written as:

$$F_R = F_{cu} - \sum_{i=1}^{k} \frac{n_i}{N_i} (F_{cu} - f_i) \tag{1.16}$$

where F_R is the residual compressive strength, F_{cu} is the compressive ultimate strength of the laminate, while f_i is the applied stress level in the ith cycle.

The predicted spectrum fatigue lives of three different laminates using the linear residual strength reduction model along with the experimental data are shown in Figure 1.32. The stress exceedance as a function of percent maximum stress for the arbitrary load spectrum used for these calculations and tests are shown in Figure 1.31. The comparison between the linear residual strength reduction model and Miner's Rule for the matrix dominated layup (16/80/4) subjected to the same spectrum is shown in Figure 1.33. This comparison shows Miner's Rule to be substantially unconservative for higher values of test limit stress.

The effects of spectra variations on the fatigue life of graphite–epoxy composite laminates was investigated by Badaliance and Dill [22]. The baseline spectrum considered was a cycle-by-cycle stress history of upper wing skin of a fighter plane. Variations of this baseline spectrum that were considered are: (1) clipping to 90% test limit stress, (2) addition of arbitrary stress overloads, (3) addition of low loads, (4) truncation to 70% test limit stress, (5) clipping of tension loads, (6) increased mission severity, i.e., increased severity and number of air-to-air loads. Exceedance curves for these spectra variations are shown in Figures 1.34, 1.35 and 1.36.

Constant amplitude fatigue data and the analytic life prediction methodology described in the foregoing was used to predict the above spectrum lives of a fiber dominated (48/48/4) and a matrix dominated (16/80/4) laminates. The specimen geometry, number of plies and stacking sequences were identical to those used in the methodology development work described in the previous sections. Predicted and test lives for the baseline spectrum are presented in Figure 1.37, where the test mean life is indicated by an arrow symbol. The value of test mean was estimated by Weibull statistical analysis. The effects of usage changes (increased severity and number of air-to-air loads) on life are shown in Figure 1.38 demonstrating a substantial decrease in the laminates lives in comparison to the baseline spectrum lives.

The results of all the spectrum variation tests and predictions are summarized in Figure 1.39. The measured and predicted lives are normalized with respect to the baseline life. The data indicates the effects of spectrum variations on fatigue life is generally the same for both the fiber (48/48/4) and matrix (16/80/4) dominated layups. The variations which have the greatest effect are those that increased the magnitude or frequency of high loads in the spectrum. It can be seen that the two variations, addition of overloads, and increased severity and number of air-to-air loads, caused more than an order of magnitude reduction in life. A 5% increase in test limit stress resulted in a 60% decrease in life, while a 5% decrease in test limit stress increases life by a factor of two. The other variation that modifies the high loads is clipping to 90% test limit stress. This variation resulted in small change in test life – an average of less than 15% increase in life. This effect is predicted with reasonable accuracy. Variations predicted to have small impact on life are those that change the lesser loads in the spectrum. These variations are the addition of low-loads to the baseline spectrum, truncation to 70% test limit stress, and clipping of tension loads.

11

Test lives for these variations are less than the predictions. Addition of low loads was found to reduce life, by a greater amount than predicted. Test data shows that removal of low loads by truncation to 70% test limit stress reduces life; however, a modest increase in life is predicted. This discrepancy of test and prediction may be caused by test scatter. Clipping of tension is predicted to have small effect on life. However, with the matrix dominated layup, test life has decreased by a substantial amount. This decrease in life is in disagreement with the fiber dominated test results and may be the result of scatter in the test data.

In general, nondelamination prone graphite—epoxy composite laminates have excellent fatigue properties. As a result, the stress levels used in the test program and in the above comparisons are much higher than would typically be used in aircraft structures. From these data it can be seen that this methodology generally overpredicts spectrum life, and that the predictions and test results are in agreement within a factor of three.

The variations in spectrum life of composite laminates discussed in the foregoing are compared in Figure 1.40 with those of metallic structure. Crack growth from open holes in .25 inch thick 7075-T7351 aluminum was used for this comparison [4]. This figure shows that there is a significant difference in the response of composite and metallic structures. It can be seen that addition of infrequent overloads tend to increase the spectrum fatigue life of metallic structures. There is a substantial reduction (approximately a factor of 50) in the life of composite laminates. The life of composite laminates tends to be less sensitive to variations in lesser loads, while the lesser loads in the spectrum significantly affect fatigue life of metals. Furthermore, the life of composite structures shows high sensitivity to change in stress level in comparison to metallic structures. For example, a 5% increase in limit stress reduces composite life by 60%, whereas the reduction in metallic structure life is approximately 10%.

1.5. References

[1]   Proceedings of the Air Force Conference on Fatigue and Fracture of Aircraft Structures and Materials, AFFDL-TR-70-144, Air Force Flight Dynamics Laboratory, Dayton, Ohio, September, 1970.

[2]   The Accumulation of Fatigue Damage in Aircraft Materials and Structures, AGARD-AG-157, Advisory Group for Aerospace Research and Development, 1972.

[3]   *Fatigue Crack Growth Under Spectrum Loads*, ASTM STP 595, American Society for Testing Materials, 1976.

[4]   Dill, H.D. and Saff, C.R., Effects of Fighter Attack Spectrum on Crack Growth, AFFDL-TR-112, Air Force Flight Dynamics Laboratory, Dayton, Ohio, November, 1976.

[5]   Roth, G.J., Development of Flight-by-Flight Fatigue Test Data from Statistical Distributions of Aircraft Stress Data, AFFDL-TR-75-16, Air Force Flight Dynamics Laboratory, Dayton, Ohio, May, 1975.

[6]   van Dijk, G.M. and de Jonge, J.B., Introduction to a Fighter Aircraft Loading Standard for Fatigue Evaluation 'FALSTAFF', NLR MP 75017, National Aerospace Laboratory, The Netherlands, May, 1975.

[7]   Schijve, J. and Jacobs, F.A., 'Program Fatigue Tests on Notched Light Alloy Specimens of 2024 and 7075 Material', National Aero Research Institute of Holland Report No. TR-M. 2070, July, 1961.

[8]   Nauman, E.C. and Schott, R.L., 'Axial Load Fatigue Tests Using Loading Schedules Based on Maneuver Loads Statistics', NASA Technical Note D-1253, May, 1962.

[9]   Breyan, W., 'Effects of Block Size, Stress Level, and Loading Sequence on Fatigue Characteristics of Aluminum Alloy Box Beams', ASTM STP 462, 970, pp. 127–166.

[10] Payne, A.O., 'Random and Programmed Load Sequence Fatigue Tests on 245T Aluminum Alloy Wings', ARL Report SM 244, September, 1956.

[11] Mayer, J.P. and Hamer, H.A., 'Applications of Power Spectral Analysis Methods to Maneuver Loads Obtained on Jet Fighter Airplanes During Service Operations', NACA RM LS6J15, January, 1957.

[12] Clementson, G.C., 'An Investigation of the Power Spectral Density of Atmospheric Turbulence', Ph.D. Thesis, M.I.T., 1959.

[13] Miles, J.W., 'An Approach to the Buffeting of Aircraft Structures by Jets', Report No. SM-14795, Douglas Aircraft Co., June 1953.

[14] Rice, S.O., 'Mathematical Analysis of Random Noise', Selected Papers on *Noise and Stochastic Processes*, edited by Nelson Wax, Dover Publication, 1954.

[15] Dill, H.D. and Young, H.T., 'Stress History Simulation', AFFDL-TR-76-113, Volume I, March 1977.

[16] Badaliance, R., 'Application of Strain Energy Density Factor to Fatigue Crack Growth Analysis', *Journal of Engineering Fracture Mechanics*, 13, No. 3, 1980.

[17] Badaliance, R., 'A Fatigue Crack Growth Theory Based on Strain Energy Density'. Proceedings of Symposium on Absorbed Specific Energy/Strain Energy Density, edited by G.C. Sih, E. Czoboly and F. Gillemot, Budapest, Hungary, September 1980.

[18] Badaliance, R., 'Mixed Mode Fatigue Crack Propagation', U.S.–Greece Symposium on 'Mixed Mode Crack Propagation', Athens, Greece, August, 1980.

[19] Sih, G.C., 'Strain Energy Density Factor Applied to Mixed Mode Crack Problems', *International Journal of Fracture*, 10, pp. 305–321, 1974.

[20] Sih, G.C., 'Strain Energy Density and Surface Layer Energy for Blunted Crack or Notches', *Stress Analysis of Notch Problem*, edited by G.C. Sih, Noordhof International Publishing, Alphen Aan Den Rijn, The Netherlands, 1978.

[21] Kipp, M.E. and Sih, G.C., 'The Strain Energy Density Failure Criterion Applied to Notched Elastic Solids', *International Journal of Solids and Structures*, 2, pp. 153–173, 1975.

[22] Badaliance, R. and Dill, H.D., 'Effects of Fighter Attack Spectrum on Composite Fatigue Life', AFWAL-TR-81-3001, March 1981.

[23] Badaliance, R. and Dill, H.D., 'Damage Mechanism and Life Prediction of Graphite/Epoxy Composites', *Damage in Composite Materials*, ASTM STP 775, 1982.

[24] Saff, C.R., 'Compression Fatigue Life Prediction Methodology for Composite Structures – Literature Survey', NADC-78201-60, June, 1980.

[25] Sih, G.C., Chen, E.P., Huang, S.L. and McQuillen, E.J., 'Material Characterization of the Fracture of Filament-Reinforced Composites', Journal of Composite Materials, 9, pp. 167–186, April 1975.

[26] Sih, G.C. and Chen, E.P., 'Fracture Analysis of Unidirectional and Angle-Ply Composites,' NADC-TR-73-1, 1973.

[27] Benthem, J.P. and Koiter, W.T., 'Asymptotic Approximation to Crack Problems', *Methods of Analysis and Solutions for Crack Problems*, edited by G.C. Sih, Noordhof International Publishing, Leyden, Volume 1, 1973.

[28] Dill, H.D. and Saff, C.R. 'Environment-Load Interaction Effects on Crack Growth', AFFDL-TR-78-137, November, 1978.

[29] Broutman, L.J. and Sahu, S., 'A New Theory to Predict Cumulative Fatigue Damage in Fiberglass Reinforced Plastics', *Composite Materials: Testing and Design (Second Conference), ASTM STP 497*, 1972, pp. 170–188.

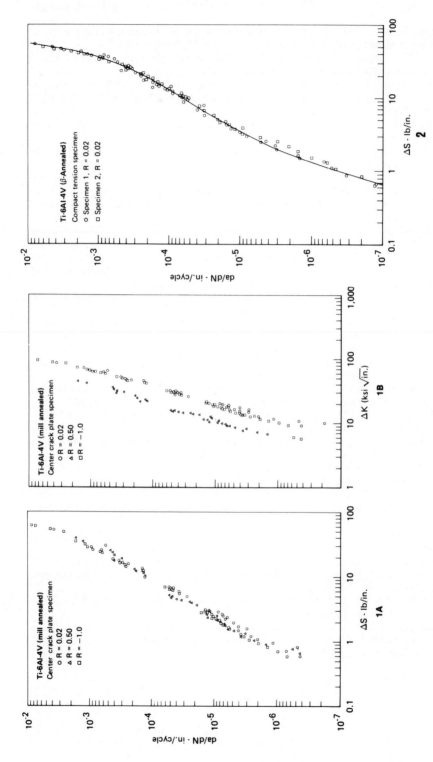

Figure 1.1. Fatigue crack growth rate of mill-annealed titanium as a function of ΔS and ΔK.
Figure 1.2. Fatigue crack growth rate of beta-annealed titanium as a function of ΔS.

14

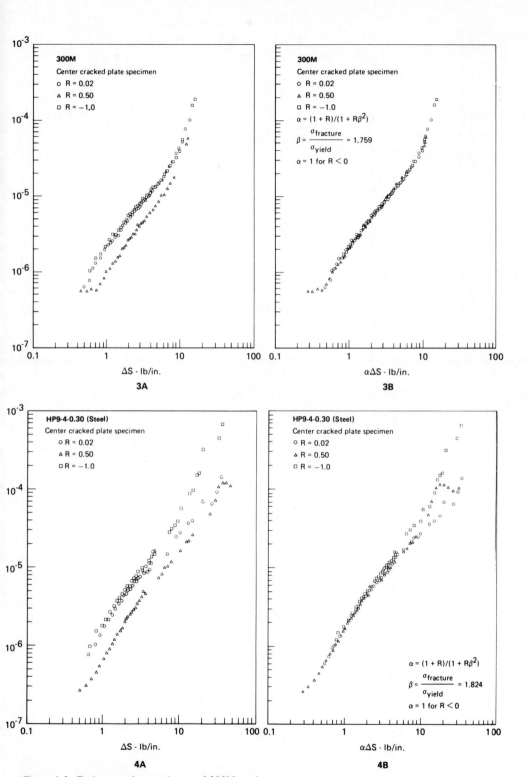

Figure 1.3. Fatigue crack growth rate of 300M steel.
Figure 1.4. Fatigue crack growth rate of HP9−4−0.30 steel.

15

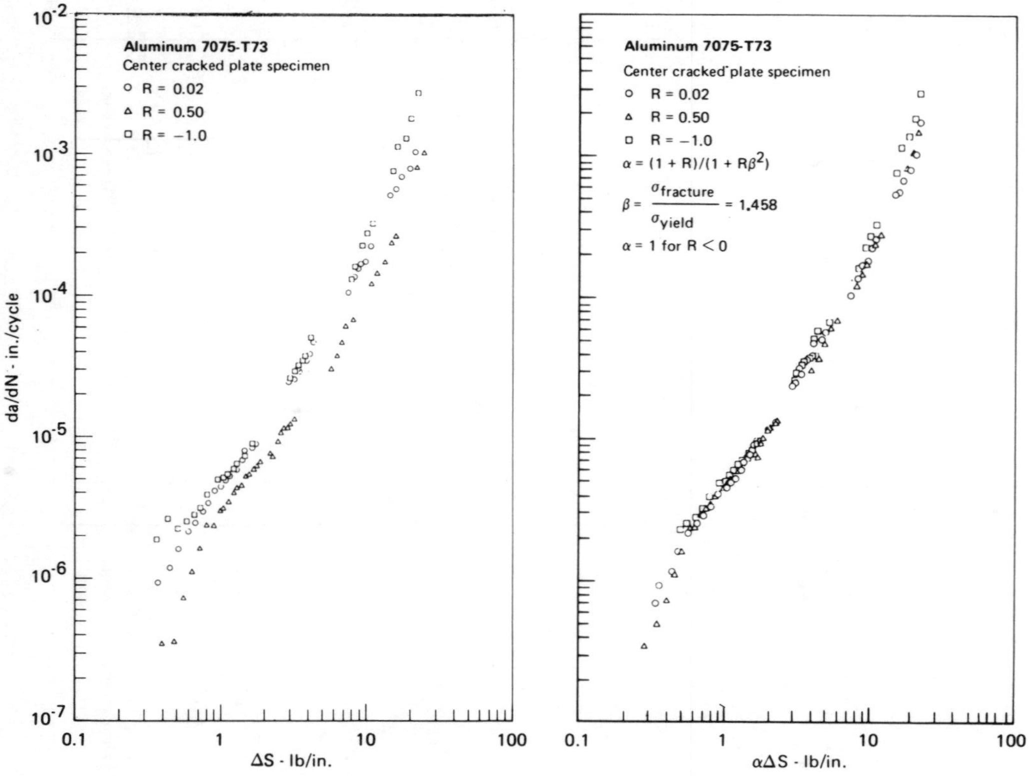

5A

da/dN - in./cycle

10^{-2}
10^{-3}
10^{-4}
10^{-5}
10^{-6}
10^{-7}

Aluminum 7075-T73
Center cracked plate specimen

o R = 0.02
▲ R = 0.50
□ R = –1.0

0.1   1   10   100

ΔS - lb/in.

5B

10^{-2}
10^{-3}
10^{-4}
10^{-5}
10^{-6}
10^{-7}

Aluminum 7075-T73
Center cracked plate specimen

o R = 0.02
▲ R = 0.50
□ R = –1.0

$\alpha = (1 + R)/(1 + R\beta^2)$

$\beta = \dfrac{\sigma_{fracture}}{\sigma_{yield}} = 1.458$

$\alpha = 1$ for R < 0

0.1   1   10   100

$\alpha \Delta S$ - lb/in.

6

da/dN - in./cycle

10^{-2}
10^{-3}
10^{-4}
10^{-5}
10^{-6}
10^{-7}

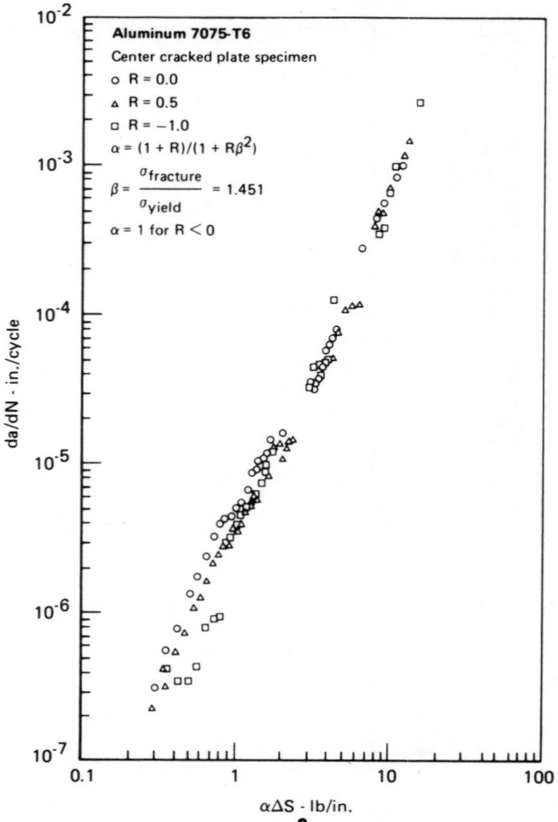

Aluminum 7075-T6
Center cracked plate specimen

o R = 0.0
▲ R = 0.5
□ R = –1.0

$\alpha = (1 + R)/(1 + R\beta^2)$

$\beta = \dfrac{\sigma_{fracture}}{\sigma_{yield}} = 1.451$

$\alpha = 1$ for R < 0

0.1   1   10   100

$\alpha \Delta S$ - lb/in.

Figure 1.5.   Fatigue crack growth rate of 7075–T73 aluminium.
Figure 1.6.   Fatigue crack growth rate of 7075–T6 aluminium.

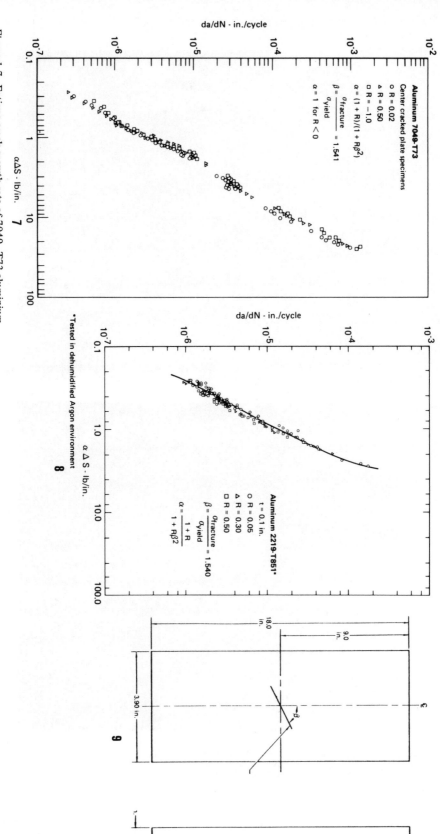

Figure 1.7 Fatigue crack growth rate of 7049–T73 aluminium.
Figure 1.8 Fatigue crack growth rate of 2219–T851 aluminium.
Figure 1.9 Inclined center crack specimen.

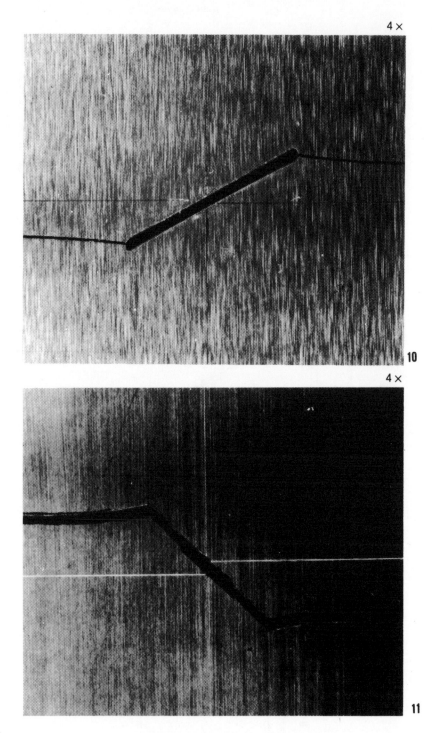

4×

10

4×

11

Figure 1.10. Appearance of precracked aluminium 7075–T7351 specimen
Figure 1.11. Appearance of 300M steel specimen.

18

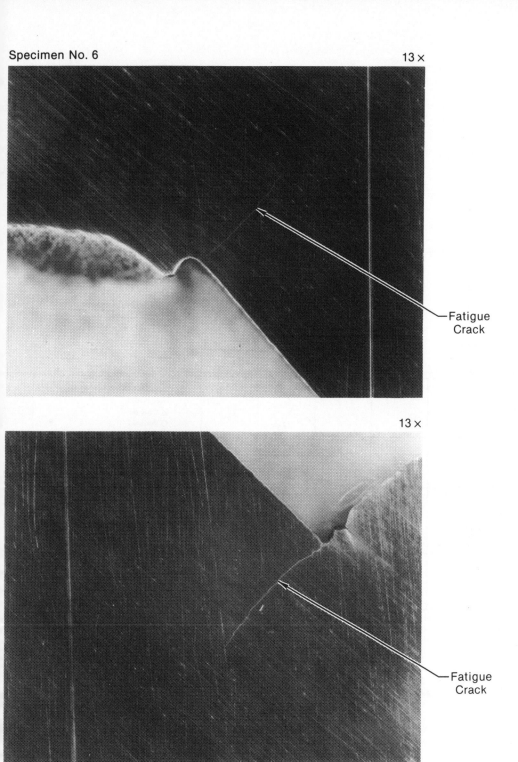

Figure 1.12. Appearance of aluminium 7075−T73 specimen subjected to compression fatigue.

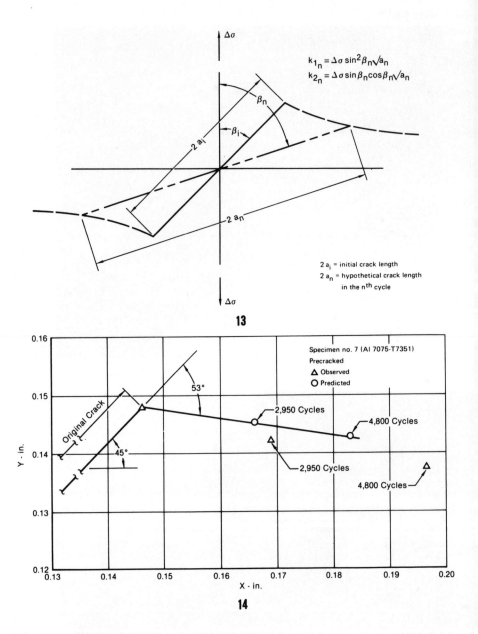

$$k_{1_n} = \Delta\sigma \sin^2\beta_n \sqrt{a_n}$$
$$k_{2_n} = \Delta\sigma \sin\beta_n \cos\beta_n \sqrt{a_n}$$

$2 a_i$ = initial crack length

$2 a_n$ = hypothetical crack length in the nth cycle

13

Specimen no. 7 (Al 7075-T7351)
Precracked
△ Observed
○ Predicted

14

Figure 1.13. Crack growth geometry approximation.
Figure 1.14. Mixed mode fatigue crack growth.

20

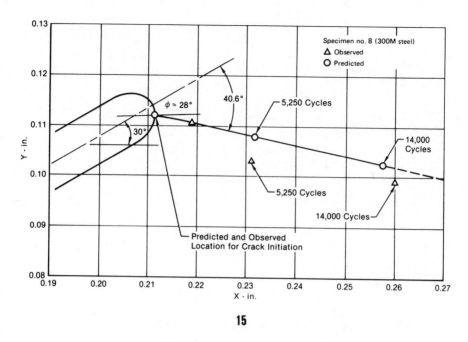

15

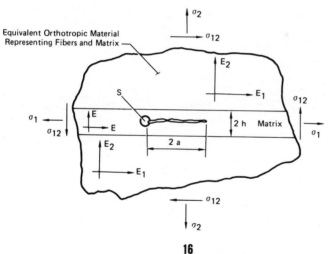

16

Figure 1.15. Predicted fatigue crack initiation location and growth from elox flaw tip.
Figure 1.16. Cracked lamina model.

21

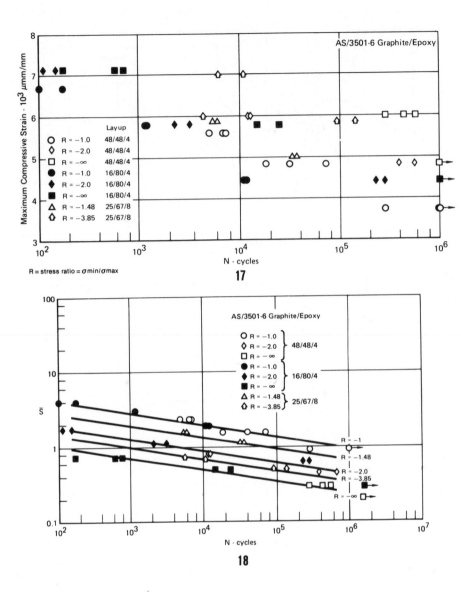

Figure 1.17. Constant amplitude fatigue results.
Figure 1.18. Correlation of S̄ with fatigue life.

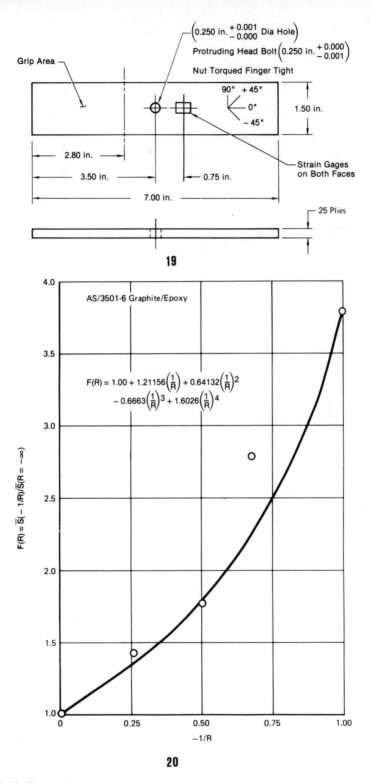

Figure 1.19. Test specimen.
Figure 1.20. Stress-ratio correction.

23

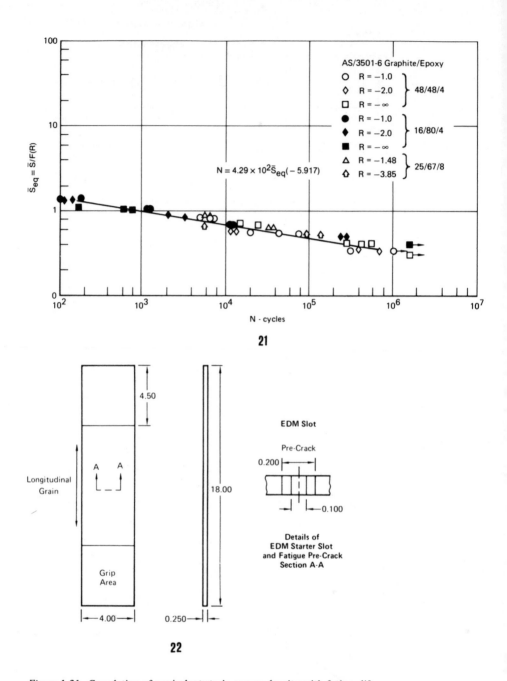

Figure 1.21. Correlation of equivalent strain energy density with fatigue life.
Figure 1.22. Center cracked plate specimen used for spectrum testing.

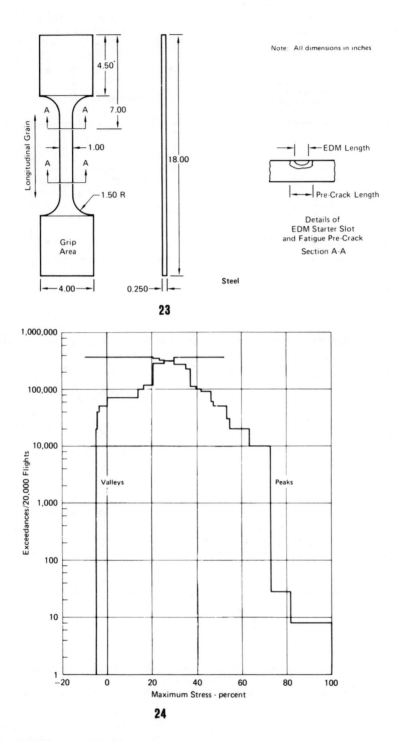

Figure 1.23. Elliptic surface flaw specimen used for spectrum testing.
Figure 1.24. Flight-by-flight stress history of a fighter plane main landing gear.

25

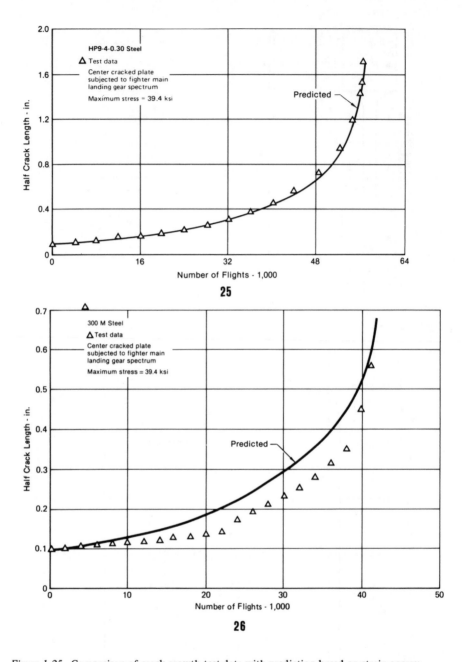

Figure 1.25. Comparison of crack growth test data with prediction based on strain energy density factor.

Figure 1.26. Comparison of crack growth test data with prediction based on strain energy density factor range (x DS).

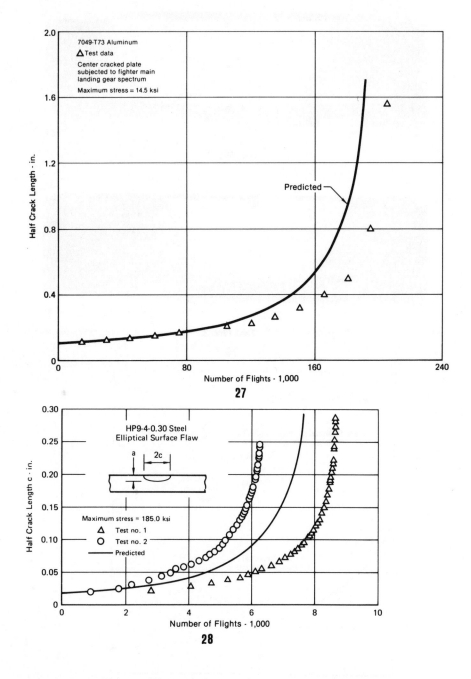

27

28

Figure 1.27. Comparison of crack growth test data with prediction based on strain energy density factor range (x DS).

Figure 1.28. Comparison of crack growth test data with prediction for surfaced flaw specimen subjected to fighter main landing gear spectrum.

4x

Test No. 1

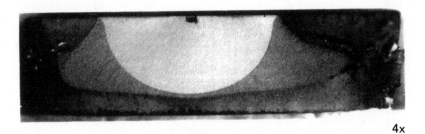

4x

Test No. 2

29

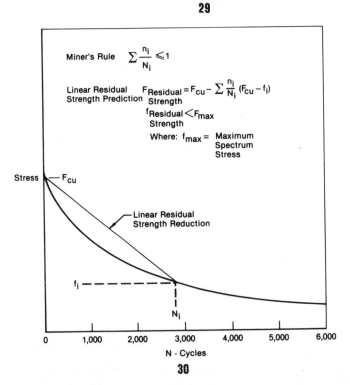

Miner's Rule $\sum \dfrac{n_i}{N_i} \leqslant 1$

Linear Residual $\quad F_{Residual} = F_{cu} - \sum \dfrac{n_i}{N_i} (F_{cu} - f_i)$
Strength Prediction Strength

$f_{Residual} < F_{max}$
Strength

Where: $f_{max} =$ Maximum
Spectrum
Stress

Stress — F_{cu}

— Linear Residual
Strength Reduction

f_i

N_i

0   1,000   2,000   3,000   4,000   5,000   6,000

N - Cycles

30

Figure 1.29. Fracture surface appearance of the HP9−4-.30 steel specimen with elliptical flaw subjected to fighter landing gear stress spectrum.
Figure 1.30. Linear fatigue damage models.

28

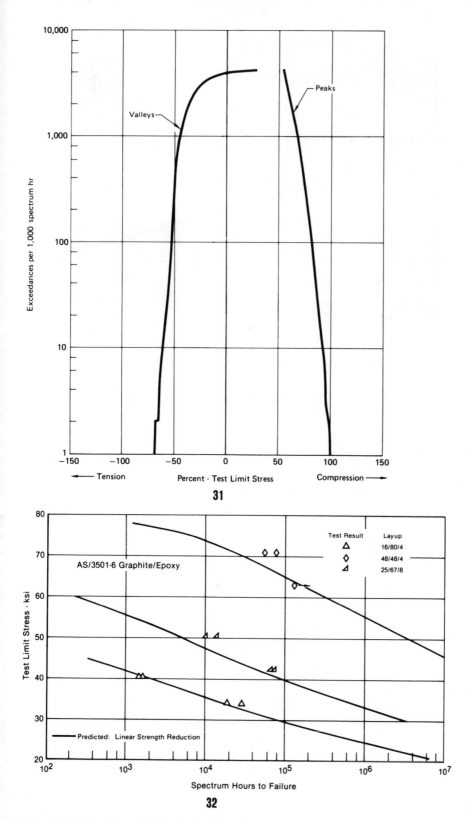

Figure 1.31. Methodology development test spectra.
Figure 1.32. Results of spectrum fatigue tests of composite laminates.

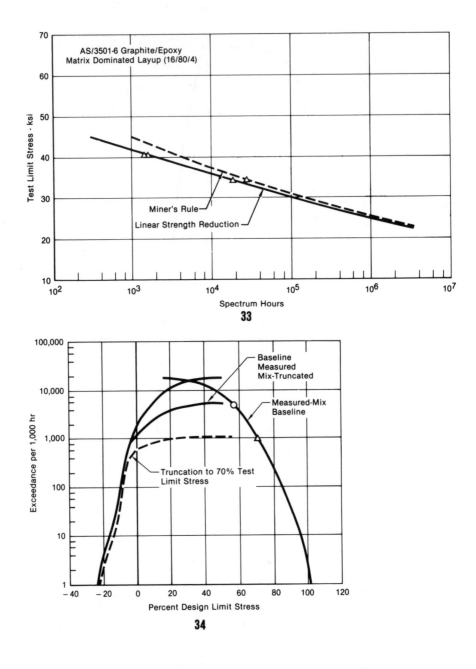

Figure 1.33. Comparison of linear strength reduction and Miner's Rule predictions.
Figure 1.34. Exceedance curves for truncation variations.

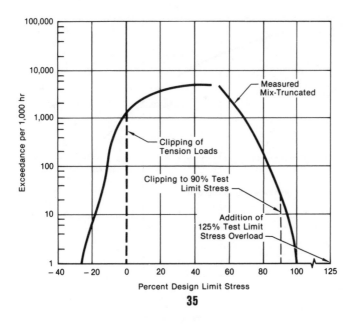

35

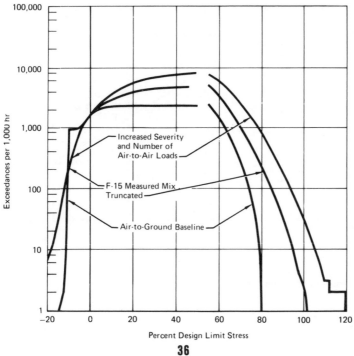

36

Figure 1.35. Exceedance curves for clipping variations.
Figure 1.36. Exceedance curves for usage variations.

31

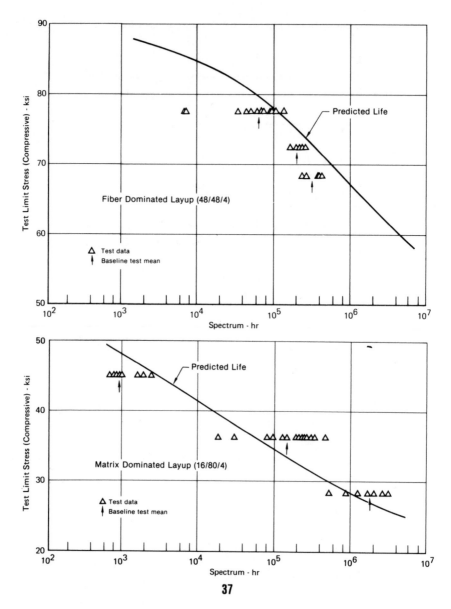

37

Figure 1.37. Baseline spectrum — measured mix, truncated.

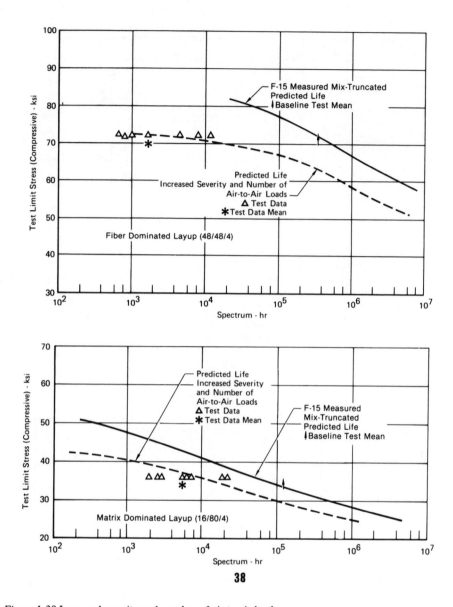

38

Figure 1.38 Increased severity and number of air-to-air loads.

33

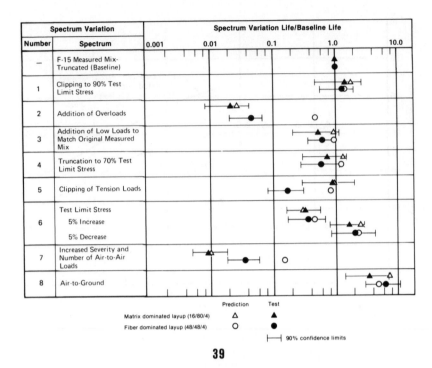

Spectrum Variation		Spectrum Variation Life/Baseline Life				
Number	Spectrum	0.001	0.01	0.1	1.0	10.0
—	F-15 Measured Mix-Truncated (Baseline)					
1	Clipping to 90% Test Limit Stress					
2	Addition of Overloads					
3	Addition of Low Loads to Match Original Measured Mix					
4	Truncation to 70% Test Limit Stress					
5	Clipping of Tension Loads					
6	Test Limit Stress, 5% Increase, 5% Decrease					
7	Increased Severity and Number of Air-to-Air Loads					
8	Air-to-Ground					

Prediction   Test

Matrix dominated layup (16/80/4)   △   ▲
Fiber dominated layup (48/48/4)   ○   ●

├——┤ 90% confidence limits

39

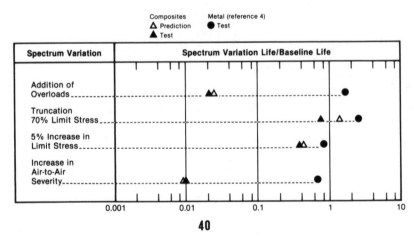

Composites   Metal (reference 4)
△ Prediction   ● Test
▲ Test

Spectrum Variation	Spectrum Variation Life/Baseline Life
Addition of Overloads	
Truncation 70% Limit Stress	
5% Increase in Limit Stress	
Increase in Air-to-Air Severity	

0.001   0.01   0.1   1   10

40

Figure 1.39. Effects of spectrum variation on life.
Figure 1.40. Effects of spectrum variations on metal and composite lives.

34

Fracture mechanics of engineering structural components

2.1. Introduction

The world-wide demand for industrialization and shortage of resources have substantially increased the need for more effective use of materials in the design of engineering structures. The technical community is faced with the responsibility of developing new and rational design procedures so that premature and/or unexpected failure can be minimized. The adoption of this seemingly obvious statement requires a philosophy and fracture criterion that have received much attention in fracture mechanics research. A major objective is to develop the capability of translating laboratory data to the design of full size structures in order to minimize cost by reducing full scale testing. It is precisely this knowledge that will ultimately determine the competitiveness of engineering products. In this respect, no technical community can afford to ignore or delay the development of fracture mechanics* technology.

Since World War II, the aircraft industry in the United States has led the efforts to understand and predict failure of metal alloys from both the material and structural engineers' viewpoints. In recent years, other industries concerned with the manufacturing of pressure vessels, pipelines, ships, bridges, etc., have also adopted the fracture mechanics discipline for much the same reasons as the aircraft industry: safety and effective material utilization. The nuclear and aircraft industries, in particular, have adopted the fracture mechanics approach for establishing inspection and nondestructive testing procedures. In view of the future stortage of metal alloys, there is also the urgent demand to strengthen the design capability of non-metallic composite materials needed for high performance structures. It is now generally accepted that conventional fracture mechanics based on K_{1c} or G_{1c} is applicable only for designing against brittle fracture, particularly in high-strength structures. In addition, the damage must be well approximated by cracks that are self-similar and aligned normal** to the applied loads. These restrictions are too severe under normal

---

*It should be clearly undertstood that fracture mechanics is not limited to the use of K_{1c} or G_{1c}. The concept deals with the characterization of material damaged by cracks. Hopefully, the so-called 'fracture toughness' parameter can be used to predict the behaviour of structural members subjected to conditions other than those simulated in the laboratory.
**Contrary to physical intuition, Mode I does not always yield the lowest critical load. Refer to the example on mixed mode crack extension in [1].

design conditions. Practical nondestructive inspection* procedures should be established on the basis of a more general theory that can predict the critical condition for different type and shape of defects located at any arbitrary position with reference to the load. Otherwise, the mere detection of defects serves no useful purpose.

In recent years, the fracture mechanics community has devoted much effort to extend the conventional theory beyond K_{1c} or G_{1c}. As research progressed in the direction of ductile fracture, however, divergent viewpoints began to spread; instead of a discipline with unifying concepts and methods, different parameters, methods of testing, and jargons emerged. The proliferation of organizations and committees within professional societies had led to the endorsement of a variety of test procedures and specimens. The variety of different specimens now being used in laboratories throughout the world make comparison of results obtained in different laboratories increasingly difficult. Moreover, the complex economic machinery of modern society exerts relentless pressure on societies such as ASTM, ASME, etc., to impose prearranged requirements on materials and design procedures in commerce. This creates the perennial danger of adopting standards and specifications before their technical implications are fully understood. The widespread feeling is that many of the activities are no longer directed at resolving the technical problem but have become self-serving and generated an exhaustive amount of unsupported data. This trend has been responsible for the paucity of progress in fracture mechanics.

This chapter deals with a brief review of the current concepts on ductile fracture and to emphasize the fundamental as well as the practical aspects of the discipline. There is no intention to propose a unified approach to fracture mechanics, rather the aim is to emphasize generality and consistency. Unless efforts are made to solve realistic engineering problems, fracture mechanics will become a subject of limited interest. It is for reasons just stated that the strain energy density criterion has been selected and used in this communication. Refer to the Introductory Chapters in [2] for application of the criterion to a host of crack problems.

2.2. Strength and fracture properties of materials

All continuum mechanics theories presuppose that the properties of the constituents can be obtained by testing a portion of the continuum with idealized geometry and loading. This is normally referred to as a specimen, the tensile specimen being the most common one. Quantities such as yield strength, fracture toughness, etc., measured from specimens are assumed to be applicable to an element or constituent of the continuum. This is accomplished conceptually by invoking failure criteria such as maximum normal stress, energy of distortion, strain energy release rate, etc. Since the specimen and continuum element may differ by more than four orders of magnitude in linear dimension, there is a size effect that is not well understood. It should also be stated at the onset that the validity of an assumed failure criterion cannot be proved mathematically or experimentally. The test of any criterion of a physical

---

*It is not sufficient to inspect only for the largest dimension of the crack or defect such as its length. The opening distance can also have a significant influence on the critical load.

nature rests on the *consistency* of its derived consequences with measurements and its usefulness in forecasting physical events. No general conclusion should be drawn from the results of a special problem. Hence, the choice of selecting a suitable failure criterion depends on the ease with which it can be applied to yield practical results.

Strain energy density function

When a solid is subjected to mechanical loading, each element will undergo deformation. With reference to a rectangular Cartesian coordinate system (x, y, z), the energy stored per unit volume ΔV, called the strain energy density function $\Delta W/\Delta V$, will be a function of (x, y, z) and possibly of time t as well if the loading is time dependent. For a linear elastic material, $\Delta W/\Delta V$ may be expressed in terms of the principal stresses σ_1, σ_2 and σ_3:

$$\frac{\Delta W}{\Delta V} = \frac{1}{2E}(\sigma_1^2 + \sigma_2^2 + \sigma_3^2) - \frac{\nu}{E}(\sigma_1\sigma_2 + \sigma_1\sigma_3 + \sigma_2\sigma_3) \tag{2.1}$$

where E is Young's modulus and ν is Poisson's ratio. It is possible to consider equation (2.1) as the sum of a part due to change of volume,

$$\left(\frac{\Delta W}{\Delta V}\right)_v = \frac{1-2\nu}{6E}(\sigma_1 + \sigma_2 + \sigma_3) \tag{2.2}$$

and of a part due to change of shape*

$$\left(\frac{\Delta W}{\Delta V}\right)_d = \frac{1+\nu}{6E}[(\sigma_1 - \sigma_2)^2 + (\sigma_2 - \sigma_3)^2 + (\sigma_3 - \sigma_1)^2] \tag{2.3}$$

such that

$$\frac{\Delta W}{\Delta V} = \left(\frac{\Delta W}{\Delta V}\right)_v + \left(\frac{\Delta W}{\Delta V}\right)_d. \tag{2.4}$$

A pictorial diagram of an element undergoing volume and shape change is shown in Figure 2.1. Generally speaking, the quantities $(\Delta W/\Delta V)_v$ and $(\Delta W/\Delta V)_d$ will vary from point to point in a continuum and their individual contribution may be related to failure by fracture and yielding.

Volume Change   Shape Change

Figure 2.1.   A continuum element undergoing volume and shape change.

*This is, in fact, the von Mises yield criterion commonly used in the continuum theory of plasticity.

Strength tests

The conventional strength test considers a cylindrical bar specimen with no visible defects or flaws. The specimen can be subjected to simple tension σ and torsion with shear stress τ as shown in Figures 2.2(a) and 2.2(b), respectively. Originally proposed by Beltrame and Haigh is a failure criterion based on the strain energy density function.

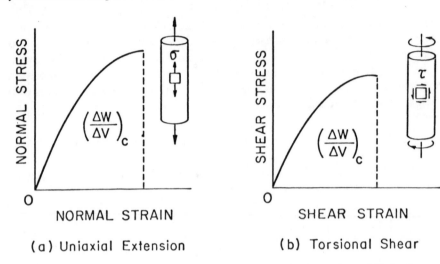

(a) Uniaxial Extension   (b) Torsional Shear

Figure 2.2.   Cylindrical bar specimen subjected to (a) simple tension and (b) torsion.

It states that failure takes place in a material when $\Delta W / \Delta V$ reaches a critical value $(\Delta W / \Delta V)_c$ being characteristic of the material. For a linear elastic material, $(\Delta W / \Delta V)$ in simple tension can be computed from equation (2.1) by setting $\sigma_1 = \sigma$ and $\sigma_2 = \sigma_3 = 0$. The result is

$$\left(\frac{\Delta W}{\Delta V}\right)_c = \frac{1}{2E}\sigma^2 \tag{2.5}$$

corresponding to the instant of failure. Similarly, in the case of torsion, $\sigma_1 = \tau$, $\sigma_2 = -\tau$ and $\sigma_3 = 0$. Equation (2.1) may again be used to give

$$\left(\frac{\Delta W}{\Delta V}\right)_c = \frac{1+\nu}{E}\tau^2 \tag{2.6}$$

at the instant of failure. Assuming that the bars in Figures 2.2(a) and 2.2(b) are made of the same material, then equations (2.5) and (2.6) may be combined to yield

$$\frac{\sigma}{\tau} = \sqrt{2(1+\nu)}. \tag{2.7}$$

As ν varies from 0.0 to 0.5, the ratio σ/τ changes from 1.414 to 1.732. This result can be checked experimentally.

The strain energy density criterion is not restricted to linear elastic materials, but

38

may also be applied to materials with nonlinear behaviour. The value $(\Delta W/\Delta V)_c$ is, in fact, the area under the true stress and strain curve and can easily be computed from experimental data [3]. Measured values of $(\Delta W/\Delta V)_c$ for a variety of metal alloys can be found in [4, 5].

Strain energy density ratio

In fracture mechanics, attention is focused on the region near the crack front where the material is severely stressed and/or strained.

Since fracture and yielding can occur simultaneously, it is necessary to consider both dilation and distortion and the corresponding energies given by equations (2.2) and (2.3). It is now well-known [6] that an element near the crack front is always in a state of plane strain* where

$$\sigma_3 = \nu(\sigma_1 + \sigma_2). \tag{2.8}$$

The stress component σ_1 is perpendicular to the crack plane while σ_2 and σ_3 are directed normal and tangent to the crack border. They form an orthogonal system. Eliminating σ_3 in equation (2.2) yields

$$\left(\frac{\Delta W}{\Delta V}\right)_v = \frac{1-2\nu}{E}\left[\sigma_1 + \sigma_2 + \nu(\sigma_1 + \sigma_2)\right]^2. \tag{2.9}$$

In the same way, equation (2.3) becomes

$$\left(\frac{\Delta W}{\Delta V}\right)_d = \frac{1+\nu}{3E}\left[\sigma_1 + \sigma_2 + \nu(\sigma_1 + \sigma_2)\right]^2 - \frac{1+\nu}{E}\left[\sigma_1\sigma_2 + \nu(\sigma_1 + \sigma_2)^2\right]. \tag{2.10}$$

It is instructive to examine the variations of the ratio $(\Delta W/\Delta V)_v/(\Delta W/\Delta V)_d$ with σ_1/σ_2. With the aid of equations (2.9) and (2.10), it is easy to see that

$$\frac{(\Delta W/\Delta V)_v}{(\Delta W/\Delta V)_d} = \frac{(1-2\nu)(1+\nu)[(\sigma_1/\sigma_2)+1]^2}{2(1+\nu)^2\left[(\sigma_1/\sigma_2)+1\right]^2 - 6(\sigma_1/\sigma_2) - 6\nu[(\sigma_1/\sigma_2 + 1)]}. \tag{2.11}$$

The numerical values of equation (2.11) for $\nu = 0.3$ are given in Table 2.1. When $\sigma_1 = \sigma_2$, $(\Delta W/\Delta V)_v$ is 6.5 times larger than $(\Delta W/\Delta V)_d$. This suggests the likelihood of fracture in an element directly ahead of the crack in tension. The ratio $(\Delta W/\Delta V)_v/(\Delta W/\Delta V)_d$ decreases as σ_1/σ_2 is increased which corresponds to more distortion. The strain energy density $(\Delta W/\Delta V)_d$ associated with shape change becomes larger than that associated with volume change when $\sigma_1/\sigma_2 \geqslant 3.36$. The gradient of $(\Delta W/\Delta V)_v/(\Delta W/\Delta V)_d$ in the thickness direction of a metal can be used to determine the amount

Table 2.1. Strain energy density ratio as a function of principal stress ratio for $\nu = 0.3$.

σ_1/σ_2	1	2	3	4	5
$(\Delta W/\Delta V)_v/(\Delta W/\Delta V)_d$	6.500	2.108	1.143	0.839	0.696

*As long as the element near the crack edge is at a finite distance away from any free surface, the condition of plane strain prevails in the interior of the solid.

of flat and slant fracture surface. A discussion with reference to pipeline design can be found in [7].

Fracture tests

As mentioned earlier, all data on materials testing refer to certain failure criterion. A unique feature of the strain energy density criterion is that it determines failure by yielding and fracture in a consistent fashion. More specifically, $\Delta W/\Delta V$ can be expressed in the form

$$\frac{\Delta W}{\Delta V} = \frac{S}{r} \tag{2.12}$$

in which S is the strain energy density factor* [2] and r the radial distance measured from the possible site of failure such as the crack tip. The singular behaviour of $1/r$ is a fundamental character of the $\Delta W/\Delta V$ criterion being independent of the constitutive equation. In other words, the $1/r$ character remains unchanged for elastic-plastic materials undergoing small or large deformation. This does not apply to criterion based on amplitude of the asymptotic stresses whose units would alter in accordance with the constitutive relation of a particular material. For instance, the order of crack tip stress singularity in the theory of finite deformation is different for each of the stress components.

The following three hypotheses on $\Delta W/\Delta V$ apply to fracture accompanied by yielding, to static and fatigue loading, and to linear as well as nonlinear materials:

Hypothesis (1). The location of fracture coincides with the location of minimum strain energy density $(\Delta W/\Delta V)_{\min}$, and yielding with maximum strain energy density, $(\Delta W/\Delta V)_{\max}$.

Hypothesis (2). Failure by stable fracture or yielding occurs when $(\Delta W/\Delta V)_{\min}$ or $(\Delta W/\Delta V)_{\max}$ reach their respective critical values.

Hypothesis (3). The amount of incremental growth $r_1, r_2, ----, r_j, ----, r_c$ is governed by

$$\left(\frac{\Delta W}{\Delta V}\right)_c = \frac{S_1}{r_1} = \frac{S_2}{r_2} = ---- = \frac{S_j}{r_j} = ---- = \frac{S_c}{r_c} = \text{const.} \tag{2.13}$$

There is unstable fracture or yielding when the critical ligament size r_c is reached.

Let a line crack in a specimen by subjected to simple tension as shown in Figure 2.3. The strain energy density $\Delta W/\Delta V$ in a local element will vary as a function of the

---

*Only for a linear elastic material that S can be computed from the stress intensity factors k_1, k_2 and k_3. Moreover, the higher order terms in the stress expressions may have to be included should the physics of the problem demand it. For notch problems, additional terms will be needed. This, however, does not constitute a change of the failure criterion.

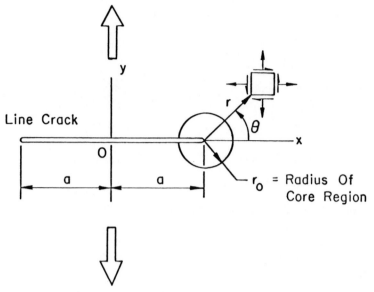

Figure 2.3. A line crack subjected to tension.

coordinates r and θ. For a fixed distance $r = r_0$, referred to as the radius* of core region, $\Delta W / \Delta V$ takes the expression

$$r_0 \left(\frac{\Delta W}{\Delta V} \right) = \frac{1 + \nu}{8E} (3 - 4\nu - \cos \theta)(1 + \cos \theta)\sigma^2 a. \tag{2.14}$$

Refer to Appendix I of Section 2.7 for a derivation of equation (2.14). The condition stated by Hypothesis (1) may be satisfied by taking the derivative of $\Delta W / \Delta V$ with respect to θ and set the result equal to zero. This gives two solutions for θ. The first, $\theta_0 = 0$, corresponds to $(\Delta W / \Delta V)_{\min}$ and determines the direction of crack initiation:

$$r_0 \left(\frac{\Delta W}{\Delta V} \right)_{\min} = \frac{(1 + \nu)(1 - 2\nu)}{2E} \sigma^2 a. \tag{2.15}$$

The second solution, $\theta_0 = \pm \cos^{-1}(1 - 2\nu)$, corresponds to locations of yielding in regions to the sides of the crack.

According to Hypothesis (2), stable fracture or damage begins as soon as $(\Delta W / \Delta V)_{\min}$ reaches the critical value $(\Delta W / \Delta V)_c$ which remains constant for a given material. The onset of rapid fracture occurs when r_0 reaches the critical ligament length, r_c, as assumed by Hypothesis (3). This is illustrated by a plot of $\Delta W / \Delta V$ versus r in Figures 2.4(a) and 2.4(b). At the onset of rapid fracture of a central crack

---

*In order to avoid the influence of microstructure, it is necessary to exclude a small region around the crack tip from the continuum analysis. The size of this zone may be estimated from a knowledge of yield strength and fracture toughness of the material [8]. Appendix II of Section 2.8 gives the results for the 4140 steel.

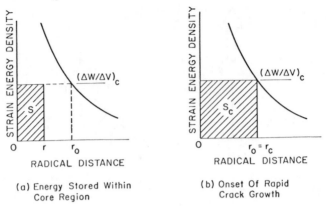

Figure 2.4.   Strain energy density factor as area under the $\Delta W/\Delta V$ versus r curve.

specimen in tension, equation (15) may be expressed in terms of the critical strain energy density factor S_c:

$$S_c = r_c \left(\frac{\Delta W}{\Delta V}\right)_c = \frac{(1 + \nu)(1 - 2\nu)K_{1c}^2}{2\pi E} \tag{2.16}$$

in which $K_{1c} = \sigma \sqrt{\pi a}$ represents the ASTM valid fracture toughness value of a particular material. A brief description on how K_{1c} is measured can be found in Appendix III of Section 2.9. Once K_{1c} is known, S_c follows from equation (2.16). Table 2.2 gives the values of K_{1c} or S_c for a number of different metal alloys with different yield strength σ_{ys}. The thickness of the specimen is denoted by h. The values of S_c in Table 2.2 are computed by assuming that $\nu = 0.33$; $E = 30 \times 10^6$ psi for steel, $\nu = 0.30$; $E = 10 \times 10^6$ psi for aluminium and $\nu = 0.33$; $E = 20 \times 10^6$ psi for titanium.

Table 2.2.   Fracture toughness values at room temperature.

Material	h (in)	σ_{ys} (ksi)	K_{1c} (ksi $\sqrt{\text{in}}$)	S_c (lb/in)
(300) Maraging Steel	0.31	242	85	17.3
D6AC Steel	0.19	217	60	8.6
AISI 4340 Steel	0.07	265	43	4.4
A533B Reactor Steel	32.4	50	180	77.7
Carbon Steel	81.6	35	200	96.0
Al 2014-T4	0.40	65	26	6.0
Al 2024-T3	0.74	57	31	8.0
Al 7075-T651	0.29	79	27	6.0
Al 7079-T651	0.49	68	30	7.4
Al DTD 5024				
Longitudinal	0.63	72	36	10.7
Thickness	0.11	70	15	1.9
Ti 6Al-4V	0.12	160	35	4.4
Ti 6Al-6V-2Sn	0.12	157	34	4.2
Ti 4Al-4Mo-2Sn-0.5Si	0.55	137	64	14.7

Compliance method

For the purpose of determining the fracture toughness value K_{1c} or S_c, it suffices to have *symmetrical* loading such that the crack is always kept normal to the applied stress. The subscript 1 on K_{1c} implies the above condition referred to as Mode I* loading. The compliance method provides a useful means of finding the stress intensity factor K_1 whose critical value is designated by K_{1c}. This method has been widely used and will be described briefly.

Let ΔW be the energy dissipated by an increment of crack extension Δa. For a plate of thickness h, the energy released by a crack extending an amount Δa is

$$G_1 = \frac{1}{2h} \frac{\Delta W}{\Delta a}. \tag{2.17}$$

Suppose that P is the load applied to a Mode I crack specimen and v is the corresponding displacement at the load-application points. For a linear elastic material, the displacement v is proportional to P. Hence, $v = CP$ with C being the compliance or the reciprocal of stiffness, $K = 1/C$. In the absence of any energy dissipation source in the material, the elastic stored energy is

$$W = \tfrac{1}{2}Pv = \tfrac{1}{2}CP^2. \tag{2.18}$$

Substituting equation (2.18) into (2.17) and maintaining the specimen at constant load, it is found that**

$$G_1 = \frac{1}{2h} P^2 \left(\frac{\Delta C}{\Delta a} \right)_{P=\text{const.}} \tag{2.19}$$

The compliance method involves measuring C in equation (19) for a series of crack lengths a from which $\Delta C/\Delta a$, the slope of the C versus a curve, can be computed. The accuracy of equation (2.19) depends on the sensitivity of the changes in the displacement between loading points remote from the crack to crack extension. This is why the method is not valid for those specimens that exhibit ductile behaviour. The quantity $\Delta C/\Delta a$ is no longer an accurate measurement of G_1 because energy loss due to plastic deformation is not being accounted for. Once equation (2.19) is known, the K_1 factor for any two-dimensional specimens, say, expressed in the form

$$K_1 = \frac{P}{bh} \sqrt{\pi a}\, f\!\left(\frac{a}{b}\right) \tag{2.20}$$

can be obtained. The dimension b is taken as the width of the specimen. A relationship between G_1 and K_1 is [9]

---

*It should be cautioned that Mode I loading does not always imply self-similar crack growth. Surface crack subjected to normal tension is an example where the crack front may change shape for each segment of growth.

**A similar expression for specimens kept at constant displacement can also be derived. The result is $G_1 = -(v^2/2h)(\Delta K/\Delta a)_{v=\text{const.}}$, where K represents the stiffness of the specimen.

$$G_1 = \frac{(1 - \nu^2)K_1^2}{E} \tag{2.21}$$

which is defined only for plane strain.* Putting equations (2.19) and (2.20) into (2.21) yields the unknown frunction f(a/b), which accounts for the specimen edge effect:

$$f\left(\frac{a}{b}\right) = b\sqrt{\frac{Eh}{2\pi a(1 - \nu^2)}} \left(\frac{\Delta C}{\Delta a}\right)_{P = \text{const.}} \tag{2.22}$$

An independent check of equation (2.22) can be made by solving for f(a/b) mathematically.

The foregoing experimental technique provides a quick way of determining K_1 or S in situations where the crack geometry is complicated and the mathematical solution may not be readily accessible.

2.3. Simple fracture experiments

The validity of the linear fracture mechanics theory may be checked by performing simple experiments on specimens with preexisting cracks. These cracks may be introduced naturally by fatigue loading as discussed in Appendix III of Section 2.9. Before delving into details, it is worthwhile to comment on the difference between the K_{1c}- and S_c-concept.

The difference between S and S_c is analogous to the difference between K_1 and K_{1c} and thus S_c is also a measure of the resistance of a material against fracture. The critical strain energy density factors S_c in Table 2.2 are particularly useful for analyzing crack systems involving more than one stress intensity factor. The additional feature of strain energy density theory or the S_c-theory is that the single parameter S_c can simultaneously determine the fracture toughness of the material and the direction of crack initiation. With this knowledge, a rational procedure of design of structural members subjected to general mode of fracture can be established. This is done by first determining S_c from a suitable test and then using S_c to set limits on the loads that can be applied to the structural member under consideration. The validity of the theory can be checked by imposing different loading conditions on the cracked specimens made of the same material and show that S_c indeed remains constant. In what follows, three basic types of loading will be discussed and they will be identified with the factors k_1, k_2 and k_3. Appendix I of Section 2.7 gives the corresponding crack tip stress fields.

Normal extension

Figure 2.5 shows a through crack of length $2a$ in a plate subjected to uniform stress σ at distances sufficiently far away from the crack. The stress intensity factor for this specimen is

*Although it is possible to write down another relation between G_1 and K_1 for the case of plane stress, it will have no meaning because the stress state near the crack border is always plane strain [10].

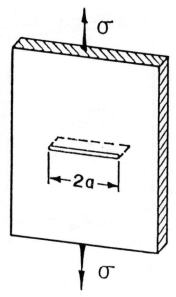

Figure 2.5. Simple tension of a through crack.

$$k_1 = \sigma \sqrt{a}, \qquad k_2 = k_3 = 0. \tag{2.23}$$

Since there exists no in-plane or out-of-plane shear, both stress intensity factors k_2 and k_3 vanish. The factor k_1 in equation (2.23) differs from K_1 in equation (2.16) by a factor of $\sqrt{\pi}$. The introduction of $\sqrt{\pi}$ in K_1 was done with no technical reason other than to absorb π into K_1^2 so that it does appear in the relationship between G_1 and K_1^2 in equation (2.21). This definition of stress intensity factor becomes unnatural in the asymptotic stresses as it would require the multiplication of $\sqrt{\pi}$ to the numerator as well as the denominator of the singular term. In order to avoid confusion, $\sqrt{\pi}$ will always enter into the fracture toughness value, K_{1c}, presented in this communication. This problem does not arise if reference is made to the critical strain energy density factor

$$S_c = \frac{(1 + \nu)(1 - 2\nu)}{2E} \sigma^2 a \tag{2.24}$$

obtained from equation (2.16).

In-plane shear

If the specimen is shear along its edges (Figure 2.6) such that the resulting stress field with reference to the crack plane is skew-symmetric, the stress intensity factors are

$$k_2 = \tau \sqrt{a}, \qquad k_1 = k_3 = 0 \tag{2.25}$$

Referring to Appendix I of Section 2.7, the strain energy density factor S is related only to k_2 through the coefficient a_{22}. In terms of the elastic properties of the specimen and the angle θ, S can be written in the form

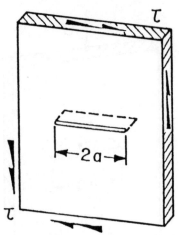

Figure 2.6.   A through crack subjected to in-plane shear.

$$S = \frac{1}{16\mu}[4(1-\nu)(1-\cos\theta) + (1+\cos\theta)(3\cos\theta - 1)]\tau\sqrt{a} \qquad (2.26)$$

where $\mu = E/2(1+\nu)$ is the shear modulus of elasticity. The direction of crack extension is found by applying Hypothesis (2) with $r = r_0 = $ const. and the result is

$$\cos\theta_0 = \frac{1-2\nu}{3}. \qquad (2.27)$$

Inserting equation (2.27) into equation (2.26) at the point of crack instability, S_c can be related to the shear stress at failure as

$$S_c = \frac{2(1-\nu)-\nu^2}{12\mu}\tau^2 a. \qquad (2.28)$$

Since S_c is assumed to be a material constant, the above expression must be equal to equation (24) for the simple tension test of the same material, i.e.,

$$S_c = \frac{2(1-\nu)-\nu^2}{12\mu}\tau^2 a = \frac{1-2\nu}{4\mu}\sigma^2 a \qquad (2.29)$$

which can be solved for the ratio

$$\frac{\sigma}{\tau} = \sqrt{\frac{2(1-\nu)-\nu^2}{3(1-2\nu)}}. \qquad (2.30)$$

For $\nu = 1/3$, equation (30) gives a theoretical prediction of $\sigma/\tau = 1.105$. The results from tests of many plexiglass plates with a central crack of length $2a$ show that on the average, this ratio is approximately 1.093, i.e., $\sigma/\tau = k_{1c}/k_{2c} = 470/430$.

Out-of-plane shear

Consider a block of linear elastic material with a through crack of width 2a whose plane lies in the longitudinal direction of the block as illustrated in Figure 2.7. The

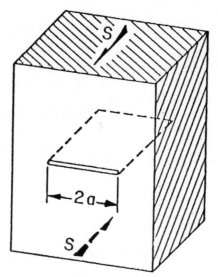

Figure 2.7.   Out-of-plane shear of a block with a through crack.

block is sheared by uniform shear stresses s acting in equal and opposite direction. Opposing shear stresses (not shown in Figure 2.7) are also applied on the front and back face of the block to maintain equilibrium. The stress intensity factors are

$$k_3 = s\sqrt{a}, \qquad k_1 = k_2 = 0. \tag{2.31}$$

From Appendix I of Section 2.7, $S = a_{33}k_3^2$ and a simple expression is found for the critical shear stress s which can be related to the critical normal stress σ in equation (2.24) by equating S_c, i.e.,

$$S_c = \frac{1}{4\mu}s^2a = \frac{1-2\nu}{4\mu}\sigma^2a \tag{2.32}$$

which yields

$$\frac{\sigma}{s} = \frac{1}{\sqrt{1-2\nu}} = 1.733 \quad \text{for} \quad \nu = 1/3. \tag{2.33}$$

Equation (2.32) may also be equated to equation (2.28) for the critical in-plane shear stress τ. The result gives

$$S_c = \frac{1}{4\mu}s^2a = \frac{2(1-\nu)-\nu^2}{12\mu}\tau^2a. \tag{2.34}$$

The ratio of the critical stresses is

$$\frac{\tau}{s} = \frac{\sqrt{3}}{\sqrt{2(1-\nu)-\nu^2}} = 1.567 \quad \text{for} \quad \nu = 1/3. \tag{2.35}$$

These predicted values of σ/s and τ/s can be verified experimentally by carrying the simple fracture tests described earlier.

Data interpretation

The interpretation of experimental data is not always straightforward, especially when the physical process of failure or fracture is not understood or unknowingly left out in the analysis. It is constructive to compare the predictions made by two different theories with the same set of experimental data.

A frequent problem is the failure of specimens under combined loading where both normal stress σ and shear stress τ are present. According to the conventional approach with no consideration given to initial defects in the material, the maximum principal stress criterion states that σ and τ at failure, when normalised by the ultimate strength σ_u takes the form

$$\left(\frac{\sigma}{\sigma_u}\right)^2 + \left(\frac{\tau}{\sigma_u}\right)^2 = 1. \tag{2.36}$$

Another criterion is the maximum total strain energy density stating that

$$\left(\frac{\sigma}{\sigma_u}\right)^2 + 2(1+\nu)\left(\frac{\tau}{\sigma_u}\right)^2 = 1 \tag{2.37}$$

at failure. There are other criteria which yield expressions similar to those given by equations (2.36) and (2.37). The choice of a particular criterion depends on how well the data would fit the assumed curve. In general, an empirical equation of the type

$$M\left(\frac{\sigma}{\sigma_u}\right)^m + N\left(\frac{\tau}{\sigma_u}\right)^n = 1 \tag{2.38}$$

is used. The parameters M, N, m and n for a particular material are determined by fitting equation (2.38) with the experimental data. The common practice is to consider these parameters as material constants and to attribute apparent changes in material behaviour. It will be shown in the subsequent example that such a procedure is unsatisfactory should the parameters be a function of the initial defects in the specimen rather than the material constants.

In order to be specific, the loading and specimen geometry in a given test are fixed but the material may or may not contain initial defects. The case of a cylindrical bar subjected to combined tension and torsion will be considered for evaluating the influence of initial defects on the parameters in equation (2.38). Let the bar of radius b in Figure 2.8 contain a penny-shaped crack of radius a. A force P and torque T are applied such that the bar experiences both a nominal normal stress $\sigma = P/\pi b^2$ and a nominal shear stress $\tau = 2Ta/\pi b^4$. The corresponding stress intensity factors are [12]

$$k_1 = \frac{\sigma}{1-(a/b)^2}\sqrt{\frac{ac}{b}}\,f_1\!\left(\frac{a}{b}\right)$$

$$\tag{2.39}$$

$$k_3 = \frac{\tau}{1-(a/b)^4}\sqrt{\frac{ac}{b}}\,f_3\!\left(\frac{a}{b}\right)$$

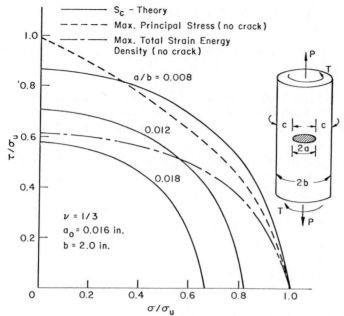

Figure 2.8.   Cylindrical bar with an internal crack subjected to tension and torsion.

in which the functions $f_1(a/b)$ and $f_3(a/b)$ stand for

$$f_1\left(\frac{a}{b}\right) = \frac{2}{\pi}\left(1 + \frac{1}{2}\frac{a}{b} - \frac{5}{8}\frac{a^2}{b^2}\right) + 0.268\frac{a^3}{b^3}$$

$$f_3\left(\frac{a}{b}\right) = \frac{4}{\pi}\left(1 + \frac{1}{2}\frac{a}{b} + \frac{3}{8}\frac{a^2}{b^2} + \frac{5}{16}\frac{a^3}{b^3} - \frac{93}{128}\frac{a^4}{b^4} + 0.038\frac{a^5}{b^5}\right).$$

(2.40)

Referring to Appendix I of Section 2.7, the strain energy density factor S for mixed mode crack extension with $k_2 = 0$ is given by

$$S = \frac{1+\nu}{8E}(3 - 4\nu - \cos\theta)(1 + \cos\theta)k_1^2 + \frac{1+\nu}{2E}k_3^2.$$

(2.41)

If $r = r_0$ is fixed, the angle θ that makes $\Delta W/\Delta V$ a minimum also gives minimum S. Therefore, by taking $\partial S/\partial\theta = 0$, letting $S_{\min} = S_c$, and using equations (2.39), it can be shown that

$$(1 - 2\nu)f_1^2\left(\frac{a}{b}\right)\sigma^2 + \frac{f_3^2(a/b)}{[1 + (a/b)^2]^2}\tau^2 = \frac{2Eb[1 - (a/b^2]^2S_c}{ac(1+\nu)}.$$

(2.42)

In view of equation (2.16), the fracture toughness S_c can be determined from a K_{1c} fracture test:

$$K_{1c} = \frac{\sigma_u}{1 - (a_0/b)^2}\sqrt{\frac{\pi a_0 c}{b}}f_1\left(\frac{a_0}{b}\right)$$

(2.43)

49

in which a_0 is the crack radius at instability. With the aid of equations (2.16) and (2.43), equation (2.42) becomes

$$\frac{f_1^2(a/b)}{f_1^2(a_0/b)}\left(\frac{\sigma}{\sigma_u}\right)^2 + \frac{[1+(a/b)^2]^{-2}}{1-2\nu}\frac{f_3^2(a/b)}{f_3^2(a_0/b)}\left(\frac{\tau}{\sigma_u}\right)^2 = \frac{a_0 c_0 [1-(a/b)^2]^2}{ac[1-(a_0/b)^2]^2}. \quad (2.44)$$

Equation (2.44) may be rearranged and cast into the form shown by equation (2.38) with $m = n = 2$. The parameter M is a complicated function of defect size and cylinder geometry and does not depend on material constants. Poisson's ratio ν is the only material constant in the expression N.

A graphical display of τ/σ_u versus σ/σ_u is given by Figure 2.8 for $a_0 = 0.016$ in, $b = 2.0$ in and a material with $\nu = 1/3$. The three solid curves labelled $a/b = 0.008$, 0.012 and 0.018 are calculated from equation (2.44) and they are plotted next to the dotted curves found from equations (2.36) and (2.37) with no consideration given to the presence of initial defects. The relation between τ/σ_u and σ/σ_u is seen to depend sensitively on the size of defects. Therefore, defects may contribute significantly to data scattering if they are not accounted for in the analysis. This illustrates the importance of interplay between theory and experiment.

2.4. Design of machine and structural components

Having decided upon the failure criterion and measured the necessary material constants, the engineer is still faced with the challenge of making use of this information to the design of machine or structural components that may acquire different shapes and subject to different loadings. Assuming that the appropriate stress analyses are performed and the stress intensity factors for the problems under consideration are known [12], it is then possible to estimate the maximum allowable load a machine or structural member can tolerate for a given defect size or the maximum defect size that is permitted to exist in a given material at the prescribed stress level. The required information depends on the design specifications.

Two types of design problems will be treated. In the first group, the loads are applied symmetrically with respect to the crack plane so that the crack, once initiated, propagates straight ahead. The direction of crack growth is known as an a priori and the K_{1c}-theory can be applied. The second group of problems considers mixed mode loading where the load may be oriented in an arbitrary location with the crack and the direction of crack growth also becomes an unknown. Under service conditions, the loads, say on an aircraft or a ship, often change directions and cannot be taken normal to the crack. To reiterate what has been emphasized earlier, symmetrical or Mode I loading does not always yield the lowest allowable load. This will be illustrated in an example to follow.

Press fit

Press fitting a shaft into the opening of another member is often encountered in engineering fracture. The presence of initial defects at the interface can significantly affect the allowable radial interference of the fit. Figure 2.9 shows a shaft of radius b

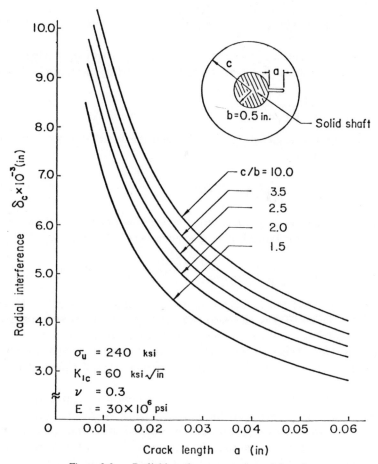

Figure 2.9. Radial interference versus crack length.

press fitted into a wheel of outside radius c. The internal diameter of the wheel is made slightly smaller than the shaft diameter so that a pressure is created when the two parts are pressed together. The maximum hoop stress along the inner boundary of the wheel is proportional to the radial interference δ and is given by [13]

$$\sigma_{max} = \frac{1}{2}E\left(\frac{\delta}{b}\right)\left[\left(\frac{b}{c}\right)^2 + 1\right] \qquad (2.45)$$

The wheel is assumed to contain a small crack of length a emanating from its inner boundary as shown in Figure 2.9. The stress intensity factor can be approximated as [12]

$$K_1 = 1.12\sigma_{max}\sqrt{\pi a} \qquad (2.46)$$

provided that the crack length a is assumed to be small in comparison with b, i.e., $a \ll b$. Eliminating σ_{max} between equations (2.45) and (2.46), a formula for finding the critical radial interference is obtained:

$$\delta_c \simeq \frac{bK_{1c}}{E\sqrt{a}} \frac{1}{1+(b/c)^2}. \tag{2.47}$$

If the wheel is made of steel with an ultimate strength $\sigma_u = 240$ ksi, fracture toughness of $K_{1c} = 60$ ksi $\sqrt{in}$, $\nu = 0.3$ and $E = 30{,}000$ ksi, equation (2.47) simplifies to

$$\delta_c \simeq \frac{10^{-3}}{[1+(b/c)^2]\sqrt{a}}, \text{ in} \tag{2.48}$$

in which $b = 0.5$ in is assumed. If $c/b = 3.5$ and the initial flaw size a $= 0.02$ in, it follows immediately that

$$\delta_c = \frac{10^{-3}}{[1+(3.5)^{-2}]\sqrt{0.02}} = 0.00654 \text{ in.} \tag{2.49}$$

The conventional maximum stress criterion which does not account for initial defects in the material predicts failure when σ_{max} in equation (2.45) reaches σ_u. This gives

$$\delta_c = \frac{2b\sigma_u}{E} \frac{1}{1+(b/c)^2} = \frac{2(0.5)(240)}{30{,}000[1+(3.5)^{-2}]} = 0.00740 \text{ in.} \tag{2.50}$$

From equations (2.49) and (2.507), a deviation of approximately 13% is observed. However, had the initial flaw size been twice as large, i.e., a $= 0.04$ in instead of a $= 0.02$ in, then equation (2.48) would render a critical radial interference $\delta_c = 0.00462$ in. In this case, the prediction based on the conventional theory will differ from that of fracture mechanics by 60%. Note that the fracture mechanics prediction is always more conservative.

Figure 2.9 gives a plot of δ_c versus crack size for $c/b = 1.5, 2.0, 2.5, 3.5$ and 10.0. As the initial flaw size is increased, the allowable radial interference must decrease in accordance with the curves of Figure 2.9 in order to avoid fracturing of the wheel during press fit. For a given flaw size, the larger wheel can sustain more interference. No gain would be made if c/b were increased indiscriminately since δ_c depends only on $1/\sqrt{a}$ for $c \gg b$.

Rotating disk

The problem of disks spinning at high speed is of vital importance in the design of steam and gas turbines as well as in many other apparatuses that involves rotating parts. Quite often the disks may contain initial cracks which may become unstable if the speed of rotation is excessive. The maximum circumferential stress of a disk with a hole occurs at the inner surface, say $r = b$. This stress increases as square of the angular velocity ω according to the expression

$$\sigma_{max} = \frac{3+\nu}{4}\rho\omega^2 c^2 \left[1 + \frac{1-\nu}{3+\nu}\left(\frac{b}{c}\right)^2\right]. \tag{2.51}$$

In equation (2.51), ρ is the mass density of the material and c is outside radius of the disk. Referring to Figure 2.10, the crack length a is assumed to be small in comparison with the hole radius b such that equation (2.46) is valid. More precisely, the ratio b/a

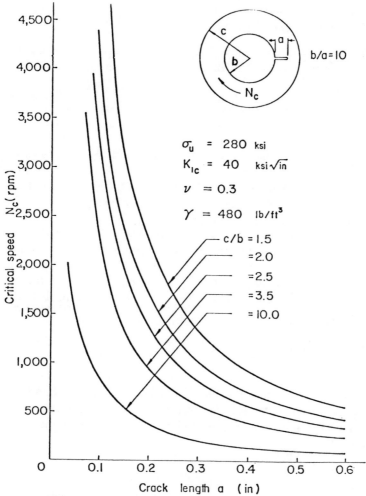

Figure 2.10. Critical speed as a function of crack length.

shall be kept at 10. Inserting equation (2.51) into (2.46) and making use of the relation $\omega = 2\pi N/60$, where N is the number of revolutions per minute the disk undergoes, an expression for the critical speed is obtained:

$$N_c = 13.560 \frac{\sqrt{K_{1c}}}{a^{1/4}\sqrt{\rho[(3 + v)c^2 + (1 - v)b^2]}}.$$ (2.52)

Once the material and geometric parameters are given for a known flaw size, equation (2.52) gives the critical speed at which the disk will burst.

A numerical example of a rotating disk made of steel with $K_{1c} = 40\,\text{ksi}\,\sqrt{\text{in}}$ and $\sigma_u = 280\,\text{ksi}$ will be considered. The mass density is $\rho = 0.104\,\text{lb/sec}^2/\text{in}^2$ and Poisson's ratio is $v = 0.3$. Consider a disk whose outside radius c is twice as large as its inside radius b and remembering that $b/a = 10$, the critical speed for an initial flaw of $a = 0.1$ is found to be

$$N_c = \frac{13.560\sqrt{40{,}000}}{(0.1)^{1/4}\sqrt{0.104[4(3.3)+0.7]}} = 4{,}020 \, \text{rpm}. \tag{2.53}$$

Based on the conventional approach, a value of N_c can be estimated directly from equation (2.51) by assuming that failure takes place when $\sigma_{max} = \sigma_u$. This leads to

$$N_c = \frac{60}{\pi}\frac{\sqrt{\sigma_u}}{\sqrt{\rho[(3+\nu)c^2+(1-\nu)b^2]}} = \frac{60\sqrt{280{,}000}}{\pi\sqrt{0.104[4(3.3)+0.7]}}$$

$$= 8{,}405 \, \text{rpm}. \tag{2.54}$$

The prediction of N_c in equation (2.53) differs appreciably from the maximum stress criterion with no consideration given to the initial flaw as calculated from equation (2.54). A difference of more than 100% is observed. The example also shows that presence of initial flaw can significantly reduce the allowable speed on rotating disks. This reduction on N_c varies approximately as $1/a^{1/4}$.

Variations of the critical speed with a variety of disk geometries and crack sizes are displayed graphically in Figure 2.10. As the size of the initial flaw increases, the speed at which the disk can rotate safely must be reduced accordingly. Note that there is a range of crack length for each ratio of c/b within which the critical speed is very sensitive to small changes in flaw size. The larger disks are particularly vulnerable to small cracks.

Thermally stressed pipe

A section of cylindrical pipe with inner radius b and outer radius c is thermally stressed due to a temperature difference ΔT across the wall. Positive ΔT indicates that the outside wall temperature is higher than the inside wall temperature. In the case of a steady state heat flow, the maximum circumferential stress at the inner surface of the pipe is

$$\sigma_{max} = \frac{\alpha E \Delta T}{2(1-\nu)}\left[\frac{2}{1-(b/c)^2} - \frac{1}{\log(c/b)}\right] \tag{2.55}$$

where α is the coefficient of linear thermal expansion of the material. Assuming that a radial flaw prevails at the inner bore, the critical temperature drop ΔT_c that causes an initial crack in the pipe to propagate can be obtained from equations (2.46) and (2.55):

$$\Delta T_c \simeq \frac{(1-\nu)K_{1c}}{\alpha E \sqrt{a}\left[\dfrac{2}{1-(b/c)^2} - \dfrac{1}{\log(c/b)}\right]} \tag{2.56}$$

Consider a steel pipe with $\sigma_u = 280 \, \text{ksi}$ and $K_{1c} = 40 \, \text{ksi}\sqrt{\text{in}}$. Use is made of the coefficient of thermal expansion $\alpha = 6.6 \times 10^{-6} \, \text{in/in/}^\circ\text{F}$, $\nu = 0.3$ and $E = 30{,}000 \, \text{ksi}$. The pipe has an outer radius to inner radius ratio of $c/b = 3/2$ and an initial flaw of $a = 0.2 \, \text{in}$. Equation (2.56) then gives

$$\Delta T_c \simeq \frac{(1-0.3)(40{,}000)}{6.6(30)\sqrt{0.2}\left[\dfrac{2}{1-0.667} - \dfrac{1}{\log(1.5)}\right]} = 281^\circ\text{F} \tag{2.57}$$

the critical temperature drop at incipient fracture. In order to avoid pipe fracture, ΔT must be kept smaller than $281°F$. Otherwise, the flaw size will have to be reduced. For example, if the initial crack length a in 0.1 in, then ΔT_c at incipient fracture is raised to $397.4°F$. Figure 2.11 gives a plot of ΔT_c versus crack length for different ratios of c/b. It is seen that the thicker pipes sustain a lower ΔT_c and are more likely to fracture for the same flaw size. Predictions made from the conventional design approach give an unrealistically high value of ΔT_c.

Stiffened panel

Present-day aircraft are protected from catastrophic fracture by restricting the cracks to spread within confined or predetermined compartments. This involves the concept of crack arrestors which are designed to reduce the crack driving force or to arrest crack motion. These crack arrestors or stringers also serve the purpose of strengthening the airframe structure and are capable of rerouting the load path.

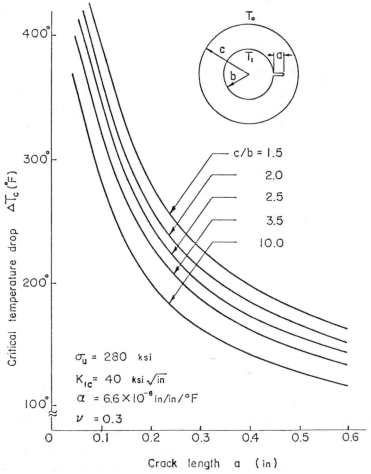

Figure 2.11. Variations of critical temperature with crack length.

For illustration, let a panel of thickness h in Figure 2.12 be stiffened by a stringer whose elastic modulus E_s is assumed to be much larger than that of the panel E, i.e., $E_s \gg E$. The panel is stretched by a uniform stress σ. As a crack of length $2a$ approaches the stringer, the stress intensity at the near end labelled (2) and the far end labelled (1) in Figure 2.12 will decrease. The normalized stress intensity factors $k_1^{(1)}/\sigma\sqrt{a}$ and $k_1^{(2)}/\sigma\sqrt{a}$ for this problem are plotted in Figure 2.13 as a function of a/b for the case of a rigid stringer [12]. The panel should be designed such that $k_1^{(2)}$ will become subcritical before the crack gets the chance to spread across the stringer. Usually, the stringers or crack arresters are spaced periodically throughout the aircraft so that $k_1^{(1)}$ will decrease even more rapidly when crack tip (1) encounters a stringer.

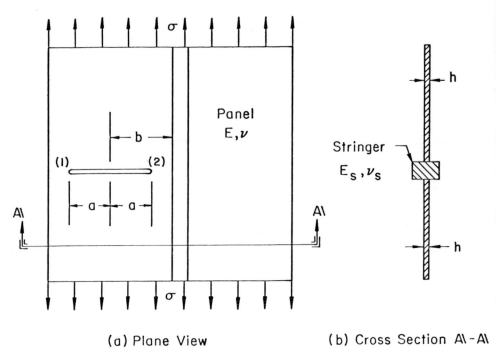

(a) Plane View   (b) Cross Section A\-A\

Figure 2.12.   A cracked plate stiffened by stringer and subjected to uniform tension.

Pressure vessels

When the stress state surrounding the crack is not symmetric with reference to the crack plane, the direction of crack initiation is no longer obvious and the K_{1c} approach cannot be used. This situation is referred to as mixed mode crack extension.

An inclined crack of length $2a$ in a pressure vessel (Figure 2.14) is an example. The angle of inclination β is referenced from the circumferential stress components σ_θ. The allowable internal pressure p and/or the limiting vessel dimension R/t will be determined for $2a = 0.2$ in and $S_c = 5.3$ lb/in for steel. The vessel radius is R and wall thickness is t. From a stress analysis, the hoop stress σ_θ and longitudinal stress σ_z are

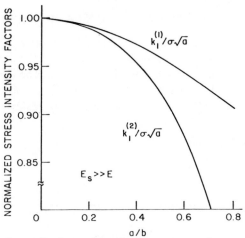

Figure 2.13. Values of normalized stress intensity factors as a crack approaches a rigid stringer.

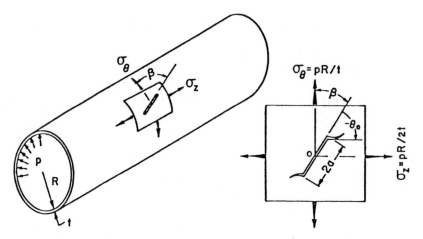

Figure 2.14. Pressurized vessel with an inclined crack.

$$\sigma_\theta = \frac{pR}{t}, \qquad \sigma_z = \frac{pR}{2t}. \tag{2.58}$$

The stress intensity factors can be found from [12]

$$k_1 = \frac{pR}{2t}\sqrt{a}(1 + \sin^2\beta), \qquad k_2 = \frac{pR}{2t}\sqrt{a}\sin\beta\cos\beta \tag{2.59}$$

which can be put into the expression for S with $k_3 = 0$. This yields

$$S = \left(\frac{pR}{2t}\right)^2 aF(\beta, \theta) \tag{2.60}$$

in which $F(\beta, \theta)$ stands for

57

$$F(\beta, \theta) = a_{11}(1 + \sin^2\beta)^2 + a_{12}(1 + \sin^2\beta)\sin^2\beta + a_{22}\sin^2\beta\cos^2\beta$$

where the coefficients a_{ij} are given in Appendix I of Section 2.7. According to the S-theory, the directions of crack propagation defined by the angles θ_0 can be found from the condition $\partial S/\partial\theta = 0$ such that S is a minimum. The values of θ_0 for $\beta = 10°$, $30°, \ldots, 90°$ and $\nu = 0.25$ are tabulated in Table 2.3. Once the values of θ_0 are inserted into equation (2.60), S becomes an intrinsic material constant S_c, i.e.,

$$S_c = \left(\frac{p_c R}{2t}\right)^2 aF(\beta, \theta_0) \qquad (2.61)$$

Table 2.3.   Fracture angles corresponding to S_{min}.

β	0°	10°	30°	50°	70°	90°
$-\theta_0$	0°	17.20°	29.36°	27.38°	17.60°	0°

from which the critical internal pressure in the vessel can be computed:

$$p_c = \frac{2t}{R\sqrt{a}}\sqrt{\frac{S_c}{F(\beta, \theta_0)}}. \qquad (2.62)$$

The numerical values of the dimensionless quantity $p_c R\sqrt{a}/2t$ for different angles of β and $\nu = 0.25$ are given in Table 2.4. If the crack is inclined at an angle $\beta = 60°$, the quantity $p_c R\sqrt{a}/2t = 12.50$ can be obtained from Table 2.4 by interpolation and the critical pressure p_c becomes

$$p_c = \frac{2}{\sqrt{a}}(12.50)\frac{t}{R} = \frac{2(12.50)}{\sqrt{0.10}}\frac{t}{R} = 79.1\frac{t}{R} \text{ ksi.} \qquad (2.63)$$

Table 2.4.   Critical vessel pressure.

β	0°	10°	30°	50°	70°	90°
$p_c R\sqrt{a}/2t$	22.50	21.60	16.95	13.54	11.80	11.30

It is apparent that once t and R for the vessel are known, p_c can be readily calculated from equation (2.63). The results in Table 2.4 also show that a longitudinal crack ($\beta = 90°$) is more dangerous as compared to a circumferential crack ($\beta = 0°$) of the same length.

Torsion of thin-walled cylinder

If the vessel in the previous example is subjected to torsion instead of internal pressure, the stress near the crack is in a state of shear. Let a through crack of length $2a$ be parallel with the axis of the cylinder as illustrated in Figure 2.15. The edges of the crack are inclined at an angle γ with the surfaces of the vessel wall. Since transverse and longitudinal shears prevail along the crack front, both stress intensity factors k_2

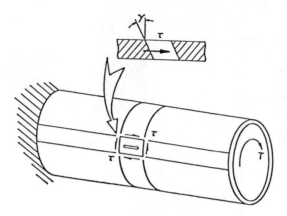

MODE II AND III

Figure 2.15.   Torsion of a cracked vessel.

and k_3 are present* while $k_1 = 0$:

$$k_2 = \tau \sqrt{a} \cos \gamma, \qquad k_3 = \tau \sqrt{a} \sin \gamma. \tag{2.64}$$

Setting $k_1 = 0$ in S and making use of equation (2.64), the strain energy density factor becomes

$$S = \tau^2 a (a_{22} \cos^2 \gamma + a_{33} \sin^2 \gamma) \tag{2.65}$$

where τ is the magnitude of the shear stress being directly proportional to the applied torque. The condition $\partial S / \partial \theta = 0$, when applied to equation (2.65), leads to the equation

$$\theta_0 = \cos^{-1} \left(\frac{1 - 2\nu}{3} \right) \tag{2.66}$$

for the direction of crack initiation. The values of θ_0 for different Poisson's ratio can be found in Table 2.5. Assuming that fracture initiates when $S_{\min} = S_c$, the critical shear stress τ_c may be found from equation (2.65) as

$$\tau_c = \frac{2 \sqrt{6 \mu S_c}}{\sqrt{a} \{2[2(1 - \nu^2) + \nu(\nu - 2)] \cos^2 \gamma + 6 \sin^2 \gamma\}^{1/2}}. \tag{2.67}$$

Table 2.5.   Fracture angles as a function of Poisson's ratio.

ν	0.0	0.1	0.2	0.3	0.4	0.5
$-\theta_0$	70.5°	74.6°	78.4°	82.3°	86.2°	90.0°

The numerical values of equation (2.67) are presented graphically in Figure 2.16. The dimensionless quantity $\tau \sqrt{a} / \sqrt{\mu S_c}$ is plotted against the crack angle γ for $\nu = 0.1$, 0.2, etc. Note that the shear stress decreases in magnitude as the angle γ increases or

*Equation (2.64) is approximate since the stress intensity factor solution does not account for the thickness effect of the vessel wall.

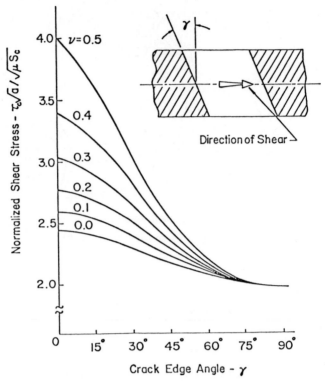

Figure 2.16.   Critical shear stress versus angle γ.

the crack edges become more slanted with respect to the vessel wall. This effect is more pronounced for materials with higher Poisson ratios.

Slanted crack in thickness direction

The plane of a through crack in laboratory specimens is often tilted at an angle α with the plate surface as shown in Figure 2.17. In this case, the state of affair near the crack border consists of a combination of normal and longitudinal shear stresses whose intensity can be described in terms of k_1 and k_3:

$$k_1 = \sigma \sqrt{a} \sin^2 \alpha, \qquad k_3 = \sigma \sqrt{a} \sin \alpha \cos \alpha \qquad (2.68)$$

where $k_2 = 0$. It follows that the strain energy density factor is

$$S = \frac{1}{16\mu} [(1 + \cos \theta)(3 - 4\nu - \cos \theta)k_1^2 + 4k_3^2]. \qquad (2.69)$$

Although this solution is approximate in that the influence of the plate surfaces is not accounted for, it can yield useful information on the failure load. The condition $\partial S / \partial \theta = 0$ predicts that the crack first runs in its own plane tilted at an angle α with the plane of loading as in Figure 2.17. This initial path will, of course, in reality, be gradually altered into a curved surface blending into a plane normal to the load.

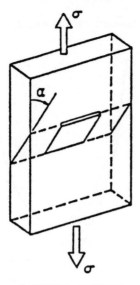

MODE I AND Ⅲ

Figure 2.17. Slanted crack.

Setting $S_{\min} = S_c$, the critical stress σ_c can be found from equation (2.69) as

$$\sigma_c = \frac{2\sqrt{\mu S_c}}{\sqrt{a}\sin\alpha\sqrt{1 - 2\nu\sin^2\alpha}}. \tag{2.70}$$

Figure 2.18 displays a plot of $\sigma\sqrt{a}/\sqrt{4\mu S_c}$ against the angle α for different values of ν. For materials with low Poisson's ratio, the applied stress tends to decrease with increasing angle α giving a minimum value at $\alpha = 90°$. However, for $\nu > 0.25$, the minimum applied stress no longer occurs at $\alpha = 90°$ but at angles determined by the equation

$$\alpha_0 = \sin^{-1}\left(\frac{1}{\sqrt{4\nu}}\right). \tag{2.71}$$

Thus, as the crack tilts away from normal to the load or Mode I loading, the specimen becomes less able to support the applied load, indicating the importance of the mixed or combined mode effect. This example shows that Mode I or K_{1c} does not always give the lowest critical applied stress.

2.5. Ductile fracture

A typical feature of ductile fracture is that the macrocrack grows slowly at first before it runs rapidly. The rate of energy release changes from slow to fast. The quantitative assessment of such behaviour requires not only a knowledge of the availability of energy for creation of free surface but also a criterion that determines the rate of incremental crack growth and shape of crack profiles which are needed for

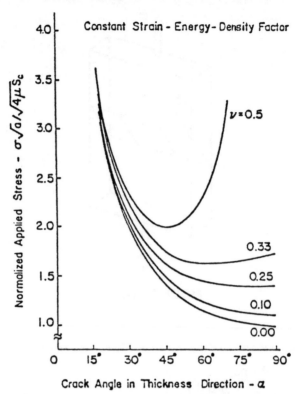

Figure 2.18.   Normalized applied stress versus crack angle α.

experimental verification. This basic transition phenomenon governs the specimen size effect that is a fundamental aspect of ductile fracture.

A brief historical account of ductile fracture is given in Appendix IV in Section 2.10 for those who wish to be informed on other possible criteria of ductile fracture. In what follows, it is essential that a clear concept of crack growth rate be established. The examples of a crack completely engulfed in a uniform stress field and a crack opened by concentrated forces applied to the crack surfaces are selected for they describe two opposing features of crack growth, i.e., one with an increasing rate and the other with a decreasing rate.

Increasing crack growth rate

Let the crack motion of the specimen in Figure 2.19(a) be regarded as a series of finite segments r_1, r_2, etc., of crack growth. The strain energy density factor S is proportional to $\sigma^2 a$, where σ is the critical stress level maintained after crack initiation. Hence, the strain energy density factors S_1, S_2, etc., will increase with each increment of growth r_1, r_2, etc., i.e.,

$$S_1 < S_2 < ---- < S_j < ---- < S_c \tag{2.72}$$

and

$$r_1 < r_2 < ---- < r_j < ---- < r_c. \tag{2.73}$$

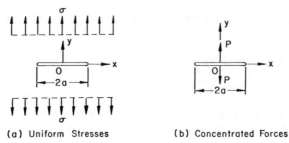

(a) Uniform Stresses   (b) Concentrated Forces

Figure 2.19.   Cracked specimens for (a) unstable crack growth and (b) stable crack growth.

It follows from the growth criterion in equation (2.13) that

$$\frac{a}{r_1} = \frac{a + r_1}{r_2} = \frac{a + r_1 + r_2}{r_3} = ---- = \text{const.} \tag{2.74}$$

If crack initiation starts when $r_1 = r_0$, then a recursion relation for incremental crack advancement can be obtained:

$$r_{n+1} = 1 + \frac{r_n}{a + r_1 + r_1 + ---- + r_{n-1}}, \qquad n \geqslant 1. \tag{2.75}$$

Each consecutive segment of growth is seen to increase and unstable fracture occurs when r_j reaches the critical ligament length, r_c. In reality, the crack front acquires a curvature across the specimen thickness with the maximum extension taking place in the mid-plane as illustrated in Figures 2.20. The growth increments $r_1, r_2, \ldots, r_j, \ldots, r_c$ in equation (2.75) correspond to the maximum values on each crack profile (Figure 2.20(a)).

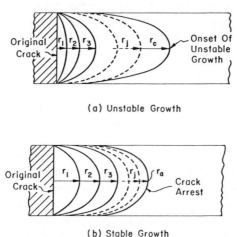

(a) Unstable Growth

(b) Stable Growth

Figure 2.20.   Incremental crack growth leading to (a) instability and (b) crack arrest.

Decreasing crack growth rate

Refer to the crack loaded by wedge forces in Figure 2.19(b). The crack begins to move by segments r_1, r_2, etc., with decreasing strain energy density S_1, S_2, etc., because S

63

is inversely proportional to the half crack length a, i.e., $S \sim P_c^2/a$ with P_c being the critical load. The conditions

$$S_1 > S_2 > ---- > S_j > ---- > S_a \tag{2.76}$$

and

$$r_1 > r_2 > ---- > r_j > ---- > r_a \tag{2.77}$$

imply that the fracture process loads to eventual crack arrest as r_j approaches r_a. Since $a(\Delta W/\Delta V)_c = \text{const.}$ during the crack growth process, equation (2.13) requires that

$$\frac{1}{ar_1} = \frac{1}{(a+r_1)r_2} = \frac{1}{(a+r_1+r_2)r_3} = ---- = \text{const.} \tag{2.78}$$

which for $r_1 = r_0$ gives the expression

$$r_{n+1} = \frac{r_n}{1 + \dfrac{r_n}{a+r_1 + ---- + r_{n-1}}}, \qquad n \geqslant 1. \tag{2.79}$$

Figure 2.20(b) shows the curved crack profiles through the specimen thickness and the decreasing increments of maximum crack growth $r_1, r_2, \ldots, r_j, \ldots, r_a$ at the mid-plane.

Construction of crack profile

A method for constructing the crack profile so frequently observed in ductile fracture will be described by application of the strain energy density criterion. The condition of constant $(\Delta W/\Delta V)_c$ will be enforced on each point of the prospective crack profile. Hence, equation (2.13) may be written in the generalized form

$$\left(\frac{\Delta W}{\Delta V}\right)_c = \frac{S_1^{(i)}}{r_1^{(i)}} = \frac{S_2^{(i)}}{r_2^{(i)}} = ---- = \frac{S_j^{(i)}}{r_j^{(i)}} = ---- = \text{const.} \tag{2.80}$$

in which $i = 1, 2, \ldots, m$ represents the number of points or elements on a crack profile and $j = 1, 2, \ldots, n$ represents the number of crack profiles in the total fracture process.

The detailed procedure for applying equation (2.80) will be presented by referring to the crack in Figure 2.21(a). The original crack front is straight and is loaded in tension. Suppose that six points or elements labelled as $1, 2, \ldots, 6$ in one-half of the specimen thickness are chosen for discussion. The locations of these elements r_1, $r_2, \ldots, r_6$ determine the crack profile from the assumption that $(\Delta W/\Delta V)_c$ is constant. Once the variations of $\Delta W/\Delta V$ with the distance measured from the original crack front are obtained either by analysis or experiment, the intersections of the line $(\Delta W/\Delta V)_c = \text{const.}$ for a material gives the values $r_1, r_2, \ldots, r_6$ as shown in Figure 2.21(b). This step-by-step procedure may be repeated until a series of crack profiles are developed from initiation to the onset of rapid fracture. A continuous record of load versus deformation of the specimen can also be made available such that a complete history of fracture and yielding is obtained.

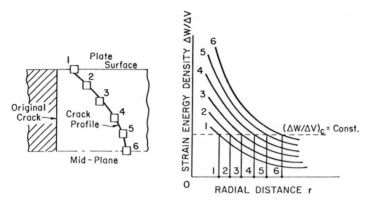

(a) Curved Crack Profile.   (b) Strain Energy Density Variations

Figure 2.21.   A procedure for constructing a curved crack front.

Ductile fracture of a through crack

Consider the example of a through crack of length $2a$ in a plate specimen of thickness h. The load is applied uniformly in a direction normal to the crack plane and is increased incrementally such that stable crack growth and plastic deformation will take place until the specimen reaches global instability. The crack profiles for each increment of growth will be predicted by incorporating the strain energy density criterion in an elastic-plastic three-dimensional finite element analysis.

For the purpose of calculation, the specimen dimensions will be specified and normalized in terms of the half-crack length a. Let the plate specimen be of height b and width c. Hence, for an initial half-crack length $a = 2.0$ in, the ratios $b/a = 4.0$, $c/a = 2.0$ and $h/a = 1.0$ are selected. Results will be found for three different values of h/a, i.e., 0.25, 0.5 and 2.0. The incremental theory of plasticity with a von Mises yield criterion is used. The material model is one of kinematic strain hardening whose uniaxial stress-strain relation takes the form

$$\epsilon = \begin{cases} \dfrac{\sigma}{E}, & \sigma \leqslant \sigma_{ys} \\[2mm] \dfrac{1}{E}\left[\sigma + \alpha\left\{\left(\dfrac{\sigma}{\sigma_{ys}}\right)^n - 1\right\}\sigma_{ys}\right], & \sigma > \sigma_{ys} \end{cases} \qquad (2.81)$$

The Young's modulus $E = 30 \times 10^6$ lb/in^2, Poisson's ratio $\nu = 0.3$ and a yield stress $\sigma_{ys} = 1.6 \times 10^5$ lb/in^2. In equation (2.81), α and n are the strain hardening parameters given as 0.2 and 2.0, respectively. For this material, the critical strain energy density is approximately $(\Delta W/\Delta V)_c = 480$ in-lb/in^3. Crack initiation is assumed to occur when $\Delta W/\Delta V$ in elements ahead of the crack front reaches the critical value.

Following the procedure described earlier in relation to Figures 3(a) and 3(b), the variations of $\Delta W/\Delta V$ with the distance r measured from the crack border are obtained numerically for $h/a = 0.25, 0.5$ and 2.0. The results are too numerous to be described here and are discussed in more detail elsewhere [14]. Only the maximum

65

crack growth increment occurring at the mid-plane will be reported as shown in Table 2.6. By keeping the increment of loading constant, the increment of crack growth r_1, r_2, etc., at the mid-plane decreases as the ratio $h/2$ is increased. The crack profiles, however, for the thinner plate are more curved than those for the thicker plate. Refer to figures 2.22 which give the crack profiles for three different plate thicknesses. The dimension d represents the depth of the plastic zone. Table 2.6 shows that the critical ligament length r_c for $h/a = 0.25$, 0.5 and 2.0 corresponds, respectively, to r_5, r_7 and r_6. This gives an average r_c value of 0.0585 in. The critical strain energy density factor for this elastic-plastic material can thus be computed as

$$S_c = r_c \times \left(\frac{\Delta W}{\Delta V}\right)_c = (5.85 \times 10^{-2} \text{ in})(480 \text{ in-lb/in}^3)$$

$$= 28.08 \text{ lb/in}.$$

Table 2.6.   Increment of crack growth in inches $\times$ 10^{-2} at mid-plane for three different plate thicknesses.

r_j \ h/a	0.25	0.50	2.00
r_1	4.60	4.25	3.40
r_2	4.70	4.35	3.80
r_3	4.80	4.55	4.20
r_4	5.10	4.80	4.70
r_5	5.90	5.05	5.05
r_6	–	5.25	5.95
r_7	–	5.70	–

The corresponding K_{1c} value is approximately $100.89 \text{ ksi} \sqrt{\text{in}}$. Note that S_c remained unchanged for different plate thicknesses. It is now clear that S_c or K_{1c} simply represents the onset of rapid fracture. Plastic deformation tends to delay global instability by enhancing slow crack growth, a process dependent on the material through the parameter $(\Delta W/\Delta V)_c$. Therefore, the complete ductile fracture process can be described by specifying any two of the critical parameters S_c and $(\Delta W/\Delta V)_c$ or S_c and r_c.

It is also of interest to observe the theoretically predicted load versus deformation relations for the three plate thicknesses considered (Figure 2.23). The deformation is interpreted as the effective displacement in the finite element analysis. The curves are nonlinear and give the critical applied loads at crack instability. The highest critical load occurred at $h/a = 0.5$. This suggests that there appears to be an optimum combination of material properties and cracked plate geometry. It must be kept in mind that improvement on the elastic-plastic stress analysis is still needed. The classical theory of plasticity and condition of yielding leave much to be desired. Nevertheless, it has been demonstrated that a fracture criterion based on the concept of strain energy density can be applied with consistency to explain ductile fracture.

Size effect

By definition, the fracture toughness parameter should be insensitive to change in specimen size such as thickness. This requirement is not being satisfied by the majority

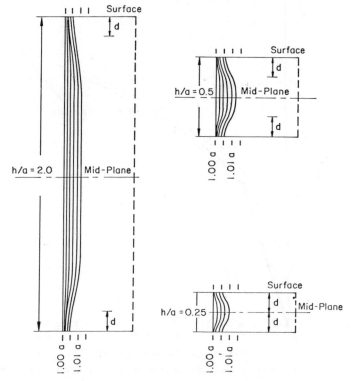

Figure 2.22.   Crack profiles for h/a ratios of 2.0, 0.5 and 0.25.

of the parameters discussed in Appendix IV of Section 2.10. It is not informative to measure different values of the parameters for different specimen thicknesses because no predictive capability will be developed. Test results of this kind are of no more value than to predict the results of other tests conducted under exactly the same conditions. This explains why the conventional material and fracture tests cannot be used to predict the performance of structures which may operate under conditions that are different from those in the laboratory.

It is fundamental knowledge that a nonlinear process is path-dependent and so must be identified with material damage at each stage of loading on the load-deformation curve. Such capability must be established theoretically and verified experimentally to resolve the size effect problem.

The specimen size effect is customarily exhibited by a plot of the critical stress σ_c times the square root of the half crack length a and π, i.e., $\sigma_c \sqrt{\pi a}$,* against the plate thickness h (Figure 2.24). Only when $\sigma_c \sqrt{\pi a}$ ceases to change for h sufficiently large is the quantity $K_{1c} = \sigma_c \sqrt{\pi a}$ a measure of the fracture toughness of the material. In this case, the excess available energy produces fracture surfaces that are predominantly flat. As h is decreased, $\sigma_c \sqrt{\pi a}$ looses its physical meaning and should no longer be

---

*Because the stress intensity factor concept is restricted to flat fracture, the measurement of K_{1c} requires the specimen to be overly thick so that a large amount of energy can be stored and released to initiate the onset of rapid crack propagation.

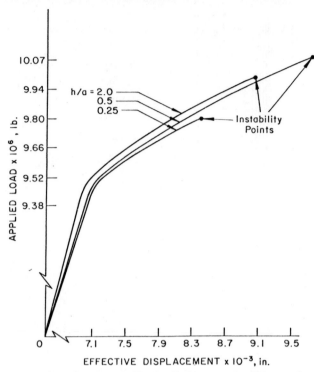

Figure 2.23.   Load as a function of effective displacement for different plate thickness.

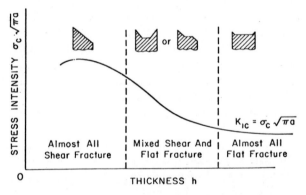

Figure 2.24.   Size effect for a cracked tensile specimen.

referred as fracture toughness or stress intensity factor. The fact that ductile fracture is accompanied by an increase in the load carrying capacity of the specimen shculd not be interpreted as a change in fracture toughness. First of all, yielding affects only the material behaviour on a load deformation diagram and does not constitute a change in material failure properties. On physical grounds, if the fracture toughness is interpreted as characterizing the resistance of the material to fracture, it should be a constant with or without yielding. Because K_{1c} is specimen size sensitive, it is more consistent to regard it simply as a measure of incipient rapid fracture rather than an inherent material parameter. Plastic deformation or yielding reduces the amount of energy available to cause unstable fracture. Ductile fracture can only be explained and resolved when the interaction between deformation and fracture are properly accounted for. The nature of this path dependent process must be understood before experiments can be designed.

An inherent characteristic of ductile fracture is that both flat and slant fracture surfaces are produced. To reiterate, Figure 2.25(a) illustrates the sequence of events in ductile fracture. The crack tends to curve during the stage of slow and stable growth while the material near the plate surfaces is deformed beyond the yield point. The local plastic deformation constrains crack growth and delays the start of unstable crack propagation. The critical strain energy density factor or fracture toughness S_c remained constant (Figure 2.25(b)) for several plate thickness to crack length ratio.

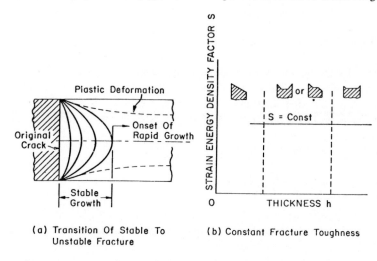

(a) Transition Of Stable To Unstable Fracture

(b) Constant Fracture Toughness

Figure 2.25. Ductile fracture in monotonically loaded tensile specimen.

As the plate thickness is continuously decreased, deformation becomes the dominant mode of failure and S_c fails to be applicable. This is why the constant S_c line has not been extended beyond the value of h below which fracture instability no longer occurs.

The transition from stable to unstable fracture is controlled by the critical ligament size r_c and fracture toughness measured by S_c which is proportional to $(\Delta W/\Delta V)_c$. Figures 2.26 give a comparison of $(\Delta W/\Delta V)_c$, S_c and critical load for two different materials A and B. Material A is said to be tougher than material B, Figure 2.26(a), if

$$\left(\frac{\Delta W}{\Delta V}\right)_A > \left(\frac{\Delta W}{\Delta V}\right)_B, \qquad \left(\frac{\Delta W}{\Delta V}\right)_A r_A > \left(\frac{\Delta W}{\Delta V}\right)_B r_B. \tag{2.82}$$

The tougher material may also sustain a higher critical load as shown in Figure 2.26(b).

The above procedure gives a complete description of deformation and fracture as the material is loaded incrementally. Each point on the load-deformation curve (Figure 2.26(b)) can be identified with material damage such that the critical load corresponds to satisfying the condition $S_c = r_c(\Delta W/\Delta V)_c$. Allowable load and net section

69

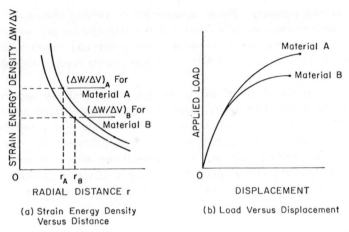

(a) Strain Energy Density
Versus Distance

(b) Load Versus Displacement

Figure 2.26.   Relationship between toughness and critical load for two materials.

size, which are the basic information needed in engineering design, can thus be found for structural members undergoing ductile fracture. The above procedure provides the analytical capability for verifying experimental results on ductile fracture. Hence, a method is now available for translating laboratory data for use in the design of larger size structural members.

2.6. Fatigue crack propagation

Despite the extensive amount of efforts made to identify fatigue damage at the earliest possible stage and to follow their subsequent growth, relatively little success has been achieved in constructing a quantitative theory. Lacking in particular is the capability to predict the fatigue life of structural members under service conditions from data collected on laboratory specimens. It is now becoming more apparent that test data alone are not sufficient unless they can be tied together by theory.

Fatigue is a path-dependent process that involves crack initiation and propagation. The necessity for addressing these two processes separately arises because the theory is not able to bridge the gap between material damage that has occurred at microscopic and macroscopic scale level. An immediate objective is to develop a procedure within the framework of continuum mechanics such that experimental data can be interpreted with order and consistency. As in the case of monotonic loading, the fatigue crack growth data should be infused with results obtained from materials testing.

Hysteresis energy density

A basic assumption of continuum mechanics is that the mechanical properties of all the elements within the material under consideration can be measured from test specimens. This notion applies to cyclic loading as well where the material element, say ahead of a crack (Figure 2.27(a)), now experiences reversal of stress and strain forming hysteresis loops as illustrated in Figure 2.27(b). Because of the multiaxial

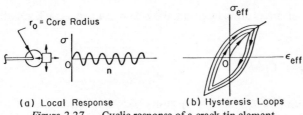

(a) Local Response   (b) Hysteresis Loops
Figure 2.27.   Cyclic response of a crack tip element.

nature of the local stress state, the effective stress σ_{eff} and strain ϵ_{eff} must be used for comparing the hysteresis energy density, $\Delta W/\Delta V$, with that obtained from a uniaxial test specimen. Since fatigue strength is load history dependent, it is not clear how data collected from constant amplitude and frequency specimens are related to the state of affairs near the crack tip region. In general, the number of cycles n in Figure 2.27(a) experienced by a local element should be distinguished from number of cycles N of the applied loading on the cracked specimen. That is, the load spectrum (Σ, N) and material response (σ, n) may not possess a one-to-one correlation.

Damage accumulation criterion

Under fatigue, the damage of the material is accumulated* on each load cycle. A convenient measure of this accumulation is the hysteresis energy density. Suppose that a saturation point will eventually be reached and the crack tip element in Figure 2.27(a) breaks allowing the macrocrack to advance. The amount of growth will depend on the rate of energy release. Let the total damage be represented in terms of the hysteresis energy density as an average,

$$\sum_{j=1}^{n}\left(\frac{\Delta W}{\Delta V}\right)_j = \left(\frac{\Delta W}{\Delta V}\right)_{ave} \Delta n \qquad (2.83)$$

where $(\Delta W/\Delta V)_{ave}$ is defined** in terms of the interval number of cycles Δn. A crack growth hypothesis can be stated:

Fatigue crack growth is assumed to occur when the total hysteresis energy density $(\Delta W/\Delta V)_{ave}\Delta n$ reaches a critical value that is characteristic of the material.

If A denotes this critical value or material constant, then the assumption becomes

$$\left(\frac{\Delta W}{\Delta V}\right)_{ave} \Delta n = A \qquad (2.84)$$

such that with the aid of equation (2.12), it can be further written as

$$\left(\frac{\Delta W}{\Delta V}\right)_{ave} = \frac{\Delta S}{\Delta r} = \frac{A}{\Delta n}. \qquad (2.85)$$

---

*The terminology "damage accumulation" should not be used to imply a monotonic relation between fatigue life and number of load cycles. An occasional high positive overload is known to increase the fatigue life of cracked specimens.
**Correction on the strain energy density for a sustained static load can be easily made.

71

If the crack advances straight ahead, then $\Delta r = \Delta a$ and a crack growth rate relation is obtained:

$$\frac{\Delta a}{\Delta N}\left(\frac{\Delta N}{\Delta n}\right) = C(\Delta S) \tag{2.86}$$

where the constant $C = 1/A$. The quantity $\Delta N/\Delta n$ cannot be set to unity unless demonstrated by analysis or experimentation. The quantity ΔS is the change of strain energy density factor for a given increment of crack growth, i.e.,

$$\Delta S = S_{j+k} - S_j. \tag{2.87}$$

In fact, ΔS is the rectangular area with sides $(\Delta W/\Delta V)_{\text{ave}}\Delta n$ or A and Δr which represents the segment of crack growth as the load is cycled from N_j to N_{j+k}, Figure 2.28. It measures the amount of irreversibility within the system. A series of ΔS values can then be computed for each increment of crack growth. This provides a rational and consistent procedure for computing the crack growth relationship, a versus N, from the basic properties of the material.

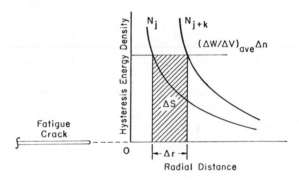

Figure 2.28.   Change of strain energy density factor in cyclic loading.

Empirical approach

Because of the difficulties associated with calculating or measuring ΔS for a real material, fatigue analysis has been approached mostly on an empirical basis [9]. The engineering approach is to perform a series of fatigue tests starting with an initial crack of approximately 0.01 in in length and measuring its growth as a function of time or number of cycles. The load is usually sinusoidal with constant amplitude and frequency, Figure 2.29. At least two of the four parameters $\Sigma_{\max}$, $\Sigma_{\min}$, $\Sigma_a(\Delta\Sigma)$ or $R = \Sigma_{\min}/\Sigma_{\max}$ are needed to define the loading. The crack growth $\Delta a/\Delta N$ can then be obtained. A variety of empirical relations have been proposed which relate $\Delta a/\Delta N$ to stress amplitude, crack length, etc. Most of the relations take the general form

$$\frac{\Delta a}{\Delta N} = B(\Delta\Sigma)^m a^q \tag{2.88}$$

where B, m and q are determined by curve fitting. There are at least more than thirty different forms of $\Delta a/\Delta N$ in the opening literature which are too numerous to be

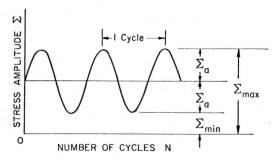

Figure 2.29. Fatigue stress components.

mentioned. The majority of them are obtained empirically. One of the most widely used equations is [15]

$$\frac{\Delta a}{\Delta N} = B(\Delta K_1)^m. \tag{2.89}$$

It was argued for aluminium alloys 2024-T3 and 7075-T6 whether the exponent m should be two (2) or four (4). The value of this exponent is, of course, not unique as $\Delta a/\Delta N$ depends sensitively on environments such as moisture or hydrogen gas and microstructure of the material. On a log $(\Delta a/\Delta N)$ versus log (ΔK_1) plot, a sigmoidal shape curve is usually obtained for metal alloys. Only the straight line portion ranging from approximately 10^{-7} to 10^{-4} in/cycle (or 10^{-6} to 10^{-3} in/cycle) can be meaningfully interpreted in terms of $\Delta a/\Delta N$. Too much significance should not be attached to the often discussed threshold stress intensity range $(\Delta K_1)_{th}$. At low values of ΔK_1, material damage occurs at a lower scale level and cannot be adequately described by the two-dimensional through crack representation of $\Delta a/\Delta N$. As the upper portion of the sigmoidal curve is approached, the crack growth rate, $\Delta a/\Delta N$, begins to rise quickly and the crack front stress intensity factor is near the critical value of K_{1c} which marks the onset of rapid fracture and terminal point of specimen or structure integrity.

The main objective in performing fatigue crack propagation studies is to be able to predict the life of structural members based on data collected from fatigue specimens. In this regard, it is pertinent that the parameters B and m in equation (2.89) should be independent of loading and specimen geometry. Figure 2.30 illustrates the end result required for predicting, say, the life of a particular structural component. The crack first grows slowly $a_1, a_2, \ldots$, until the useful life of the structural component is reached. The crack then begins to accelerate and propagate very rapidly causing the material to fragment. Initial cracks in materials are assumed to exist or to occur at an early stage of fatigue life. The form of equation (2.89), however, is not adequate for two reasons aside from the fact that it contains no mechanism of damage accumulation. First, any crack growth expression, $\Delta a/\Delta N$, should contain, in principle, at least two loading parameters, say the stress amplitude, $\Delta\Sigma$, and the mean stress ratio R so that the fatigue loading is properly defined. Equation (2.89) involves $\Delta\Sigma$ only and is restricted to crack running straight ahead. In practice, the direction of the applied load may change and it would be overly restricted to assume that the load and

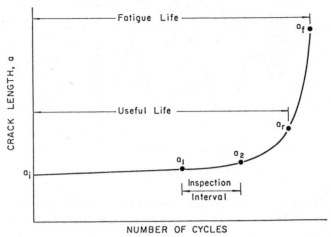

Figure 2.30.   Crack length versus number of cycles curve for constant amplitude loading.

crack should always be maintained normal to one another. The majority of the failure in service is of the mixed mode type where the crack does not propagate in the direction normal to the applied load because of the lack of geometric symmetry. Such an effect must be accounted for in the fatigue crack growth prediction analysis and has been shown to have a significant contribution for cracks initiating under monotonic loading [2].

Linearized crack growth rate relation

As mentioned before, the load spectrum (Σ, N) and local response (σ, n) of the material ahead of a fatigue crack may not be linearly related and hence the quantity $\Delta N/\Delta n$ in equation (2.86) may not necessarily be unity for a dissipative system. No realistic constitutive equations exist for evaluating the phase shift and amplitude change between the cyclic load input and local material response. The continuum theory of plasticity can provide some information [16] but it does not adequately model the physical damage of the material.

Realizing the difficulties associated with evaluating ΔS for the process of damage accumulation, an empirical approach is often adopted by collating experimental data in the form

$$\frac{\Delta a}{\Delta N} = C(\Delta S)^n \tag{2.90}$$

in which the change of the strain energy density factor ΔS is calculated from the linear theory of elasticity. In this case, $\Delta S = S_{max} - S_{min}$ with S_{max} and S_{min} corresponding respectively to Σ_{max} and Σ_{min} in Figure 2.29. Typical plots of log $\Delta a/\Delta N$ versus log ΔS for 300M steel, 7075-T6 aluminium and Ti-6Al-4V titanium are shown in Figures 2.31 to 2.33 for three different stress ratios $R = 0.02$, 0.50 and -1.0. Except for the aluminium material, all data points lie nearly on a straight line.* The

---

*The data points would be badly scattered had log $\Delta a/\Delta N$ been plotted against log ΔK_1.

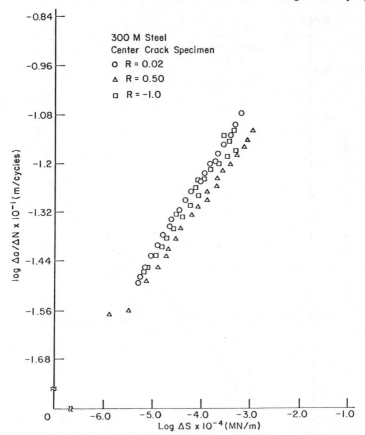

Figure 2.31. Crack growth rate versus change of strain energy density factor for 300M steel.

parameter C represents the y-intercept and n the slope of the lines drawn through the data points in Figures 2.31 to 2.33. The results are summarized in Table 2.7.

For cracks that do not grow straight ahead, Hypotheses (1)* and (2) stated earlier for monotonic loading can also be extended to the case of cyclic loading by assuming that the crack still grows in the direction of S_{min} and hence ΔS in equation (2.90) will be replaced by S_{min}, i.e.,

$$\frac{\Delta a}{\Delta N} = C(\Delta S_{min})^n. \tag{2.91}$$

More specifically, ΔS_{min} stands for

$$\Delta S_{min} = S_{min}^{max} - S_{min}^{min} \tag{2.92}$$

where S_{min}^{max} and S_{min}^{min} are the maximum and the minimum values of the strain energy density factor in the direction $\theta = \theta_0$, i.e.,

$$S_{min}^{max} = S(\theta_0, \sigma_{max}), \qquad S_{min}^{min} = S(\theta_0, \sigma_{min}) \tag{2.93}$$

*Preliminary tests at Lehigh University have shown that the direction of crack growth for monotonic and cyclic loading is approximately the same.

75

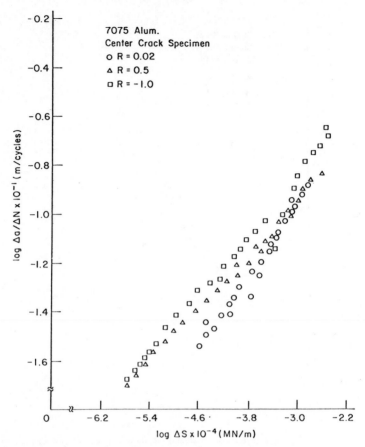

Figure 2.32. Crack growth rate versus change of strain energy density factor for 7075-T6 aluminium

which applies for a crack that can propagate on a curved path. The change of strain energy density factor in terms of the stress intensity factors takes the form

$$\Delta S_{min} = a_{11}(\theta_0)[(k_1)^2_{max} - (k_1)^2_{min}] + 2a_{12}(\theta_0)[(k_1)_{max}(k_2)_{max}$$
$$- (k_1)_{min}(k_2)_{min}] + a_{22}(\theta_0)[(k_2)^2_{max} - (k_2)^2_{min}]$$
$$+ a_{33}(\theta_0)[(k_3)^2_{max} - (k_3)^2_{min}].$$

(2.94)

Defining the stress intensity factor ranges Δk_j and the mean stress intensity factors $\bar{k}_j (j = 1, 2, 3)$, equation (2.94) becomes

$$\Delta S_{min} = 2[a_{11}(\theta_0)\bar{k}_1 \Delta k_1 + a_{12}(\theta_0)(\bar{k}_2 \Delta k_1 + \bar{k}_1 \Delta k_2) + a_{22}(\theta_0)\bar{k}_2 \Delta k_2$$
$$+ a_{33}(\theta_0)\bar{k}_3 \Delta k_3].$$

(2.95)

It should be pointed out that equation (2.95) includes not only the stress range but also the mean stress. Once the factors k_j are known for monotonic loading, ΔS_{min} may be calculated to determine the crack growth rate. A number of example problems will be considered.

76

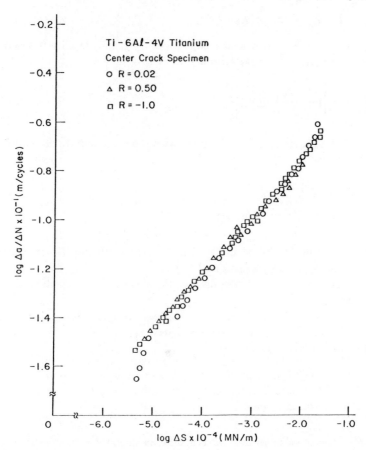

Figure 2.33. Crack growth rate versus change of strain energy density factor for Ti-6Al-4V titanium.

Table 2.7. Material and fatigue parameters for three different metal alloys in equation (2.90).

Material	ν	$E\ (10^5)$ MPa	σ_{ys} MPa	K_{1c} MPa $\sqrt{m}$	C (m/cycle)(MN/m)$^{-n}$	n
Steel (300M)	0.3	2.02	1,620	56.92	1.656×10^{-3}	1.234
Aluminium (7075-T6)	0.33	0.65	538	32.97	1.148×10	2.188
Titanium (Ti-6Al-4V)	0.31	1.28	917	71.53	7.391×10^{-1}	1.901

Center crack fatigue specimen

As a simple example, the number of cycles to failure of a center cracked specimen will be computed by using equation (2.91). For a crack of length 2a that is loaded symmetrically, the stress intensity factors for this problem are

77

$$k_1 = \Sigma \sqrt{a}, \qquad k_2 = k_3 = 0. \tag{2.96}$$

The direction of crack growth is first determined by taking $\partial S/\partial \theta = 0$ which leads to $\theta_0 = 0$. This gives $a_{11} = (1 - 2v)/4\mu$ and

$$\frac{\Delta a}{\Delta N} = C \left[\left(\frac{1 - 2v}{2} \right) a \Delta \Sigma \bar{\Sigma} \right]^m \tag{2.97}$$

in which $2\bar{\Sigma} = \Sigma_{max} + \Sigma_{min}$ is defined as twice the mean applied stress. Integration leads to the result

$$N_f = \frac{a_f^{1-n} - a_i^{1-n}}{C(1-n) \left[\dfrac{1-2v}{2} \Delta \Sigma \bar{\Sigma} \right]^n} \tag{2.98}$$

with a_i and a_f being the initial and final half crack length and N_f corresponds to a_f determined from S_c given by equation (2.16).

It is interesting to note that regardless of the mean stress $\bar{\Sigma}$, equation (2.89) yields the same value of N_f:

$$N_f = \frac{2}{B(2-m)(\Delta \Sigma)^m} \left[a_f^{1-(m/2)} - a_i^{1-(m/2)} \right]. \tag{2.99}$$

For the same data [17], the parameters B and m are different from those tabulated in Table 2.7 for the $\Delta a/\Delta N$ versus ΔS model. Table 2.6 gives the values of B and m for the $\Delta a/\Delta N$ versus ΔK_1 model.

Table 2.8.   Material and fatigue parameters in equation (2.89) for three different metal alloys.

Material	B $(\text{m/cycle})(\text{MPa/m})^{-m}$	m
Steel (300M)	2.889×10^{-11}	2.647
Aluminium (7075-T6)	5.057×10^{-12}	4.347
Titanium (Ti-6Al-4V)	4.185×10^{-12}	3.809

The predictive capability of equations (2.98) and (2.99) may be examined. Based on the parameters in Table 2.7 and 2.8 which were evaluated from data [17] for $R = 0.02, 0.50$ and -1.00, the a versus N relationshops may be calculated for another mean stress level, say $\Sigma_{max} = 210 \, \text{MPa}$ and $\Sigma_{min} = 0$ with $R = 0$, a condition that was not tested. Figure 2.34 shows that the predictions from equations (2.98) and (2.99) are substantially different. The solid curves correspond to the ΔS model and dotted curves to the ΔK_1 model which is much more sensitive to changes in the mean stress or R value. The solid curves are closer to the original data which were obtained for R other than zero.

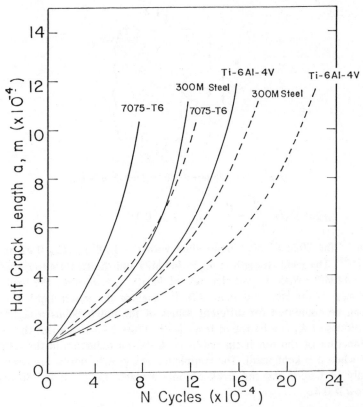

Figure 2.34. Fatigue life prediction of center crack panel with $\Sigma_{max} = 210\,\text{MPa}$ and $\Sigma_{min} = 0$ for three different metal alloys.

Effect of edge notch in fatigue

It is well-known that notches can greatly influence the fatigue life of structural members. Cracks are often found to initiate from notch sites. Consider the case of an external elliptical notch with depth d and opening $2b$ as shown in Figure 2.35. The notch radius of curvature is $\rho = b^2/d$. A crack of length a is initiated owing to the reversal loading σ applied at distances far away from the notch.

For this problem [18], the magnitude of the stress intensity factor,* k_1, will depend on the notch curvature radius ρ and the stress concentration factor K_t as follows

$$k_1 = 1.1215\Sigma\sqrt{a}\,K_t\left[\frac{1 + (a/\rho)^2}{\left(1 + \dfrac{a}{\rho}\right)^2}\right] \quad \text{for} \quad a < 0.4 \tag{2.100}$$

and for longer cracks when ρ and K_t no longer affect the notch front stresses, k_1 takes the simpler form

*Equations (2.100) and (2.101) were found to agree well with the known results obtained numerically and they serve as good approximations to the edge notch problem.

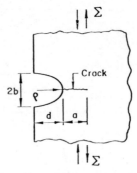

Figure 2.35.   External elliptical notch in fatigue.

$$k_1 = 1.1215\Sigma \sqrt{a} \sqrt{1 + \frac{d}{a}} \quad \text{for} \quad a > 0.4. \tag{2.101}$$

Referring to the 2024-T3 aluminium alloy with $n = 1.58$ and $C = 0.02$ (m/cycle) $\times$ $(MN/m)^{-1.58}$. The yield strength is $\sigma_{ys} = 345$ MPa and the fracture toughness is given by $K_{1c} = 44$ MPa. With an initial crack of length $a_i = 0.1$ mm, the load oscillates between $\sigma_{max} = 150$ MPa and $\sigma_{min} = 0$. By keeping the notch depth d at 10 mm, ΔS_{min} can be computed for different values of the notch radius ρ and the corresponding values of K_t can be found from [19]. Table 2.9 summarizes the fatigue life N as a function of the notch tip radius ρ. A drastic reduction in the fatigue life is observed when ρ is kept small. The complete crack growth results are given in Figure 2.36. Only in Case 1 did the crack become unstable for $a < 0.4\rho$. All other cases pertain to $a > 0.4\rho$.

Table 2.9.   Number of cycles to failure for an external elliptical notch.

Case	Notch radius ρ (mm)	K_t	Fatigue life N
1	0.2	15.5	61,900
2	1.0	7.4	65,200
3	2.0	5.5	72,000
4	5.0	3.8	97,000
5	8.0	3.25	126,900
6	10.0	3.0	136,600
7	50.0	1.9	367,000

Mixed mode fatigue crack propagation

When the fatigue load is not normal to the crack plane, the direction of crack propagation is an unknown in the problem. The case of an inclined crack is illustrated in Figure 2.37. Here, the angle $-\theta_0$ for each incremental growth of the crack must be known before the point of crack instability can be determined. Prior to the first increment of growth, the stress intensity factors k_1 and k_2 corresponding to an initial crack length of $2a_0$ are known as follows [12]

$$k_1 = \Sigma \sqrt{a_0}\sin^2\beta_0, \qquad k_2 = \Sigma \sqrt{a_0}\sin\beta_0\cos\beta_0. \tag{2.102}$$

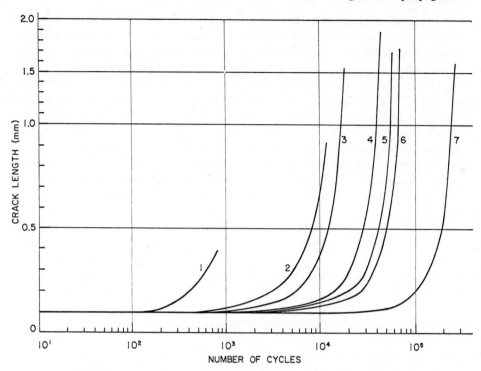

Figure 2.36.   Crack growth prediction for an external elliptical notch.

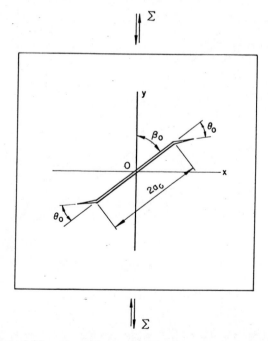

Figure 2.37.   Reversal extensional loading on an inclined crack.

81

The angles of crack initiation have been computed and are given in Figure 2.38 as a function of β for different values of the Poisson's ratio ν. Since these results are based on equations (2.102), they are valid only for the first increment of crack growth. Once the crack has adopted a zig-zag pattern, the expressions of k_1 and k_2 are altered and cannot be easily determined.* For the purpose of illustrating the method of solution based on the strain energy density model, an approximate procedure for estimating k_1 and k_2 will be used. Consider a small increment of crack growth from A to B in Figure 2.39(a) such that the newly developed bent crack OAB may be approximated by the straight line OB. Because of skew-symmetry, it is only necessary to discuss the development for one-half of the crack geometry. Thus, the new crack angle is $\beta_1 = \beta + \Delta\beta$, corresponding to the half crack length $a_1 = a_0 + \overline{CB}$ as shown in Figure 2.39(b). It follows from a geometric consideration that

$$\beta_1 = \beta + \frac{\Delta a_1 \sin \theta_0}{a_0 + \Delta a_1 \cos \theta_0} \tag{2.103}$$

and

$$a_1 = a_0 + \frac{\Delta a_1 + a_0 \cos \theta_0}{a_0 + \Delta a_1 \cos \theta_0} \Delta a_1. \tag{2.104}$$

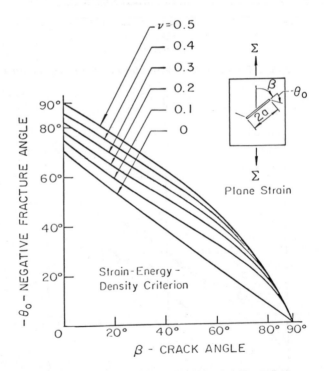

Figure 2.38. Crack angle versus fracture angle in tension.

---

*Numerical computations are also difficult as the accuracy of k_1 and k_2 will depend on the arbitrarily choice of the crack growth segments Δa_1, Δa_2, etc., and the sharpness of the bend.

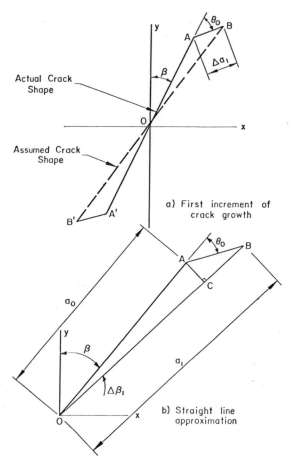

Actual Crack
Shape

Assumed Crack
Shape

a) First increment of
crack growth

b) Straight line
approximation

Figure 2.39.   Incremental crack growth initiating from an inclined crack; (a) First increment of
crack growth; (b) Straight line approximation.

Under these considerations equations (2.102) remain valid once a_0 and β_0 in equations
(2.103) and (2.104) are replaced by a_1 and β_1. By means of equation (2.94) with $k_3 =$
0, ΔS_{min} for each increment of crack growth may be computed and the number of
cycles N_1, N_2, etc., for each increment of crack growth $\Delta a_1, \Delta a_2$, etc., can be calculated.

An actual example of the fatigue of an inclined crack will now be analyzed. The
loading and crack geometry data are given in Table 2.10 for a specimen made of Ti-
6Al-4V mill annealed titanium whose fatigue properties are $C = 0.412$ and $n = 1.87$.
The analytical results pertaining to the fatigue crack trajectories and the number of
cycles for each increment of crack growth are summarized in Figures 2.40 and 2.41
and compared with the available experimental data [20]. It is seen that the predicted
crack path represented by the dotted curves are in close agreement with the actual
crack trajectories drawn by the solid lines. The number of cycles N_1, N_2 and N_3 for
the three increments of crack growth deviated appreciably from the experimental
data for the case of $\beta = 30°$ in Figure 2.40 while the analytical findings agreed reasonably well with experiments for $\beta = 43°$ in Figure 2.41. The discrepancy between the

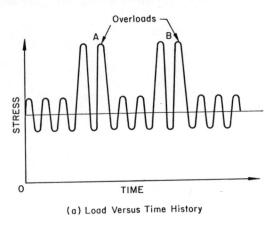

(a) Load Versus Time History

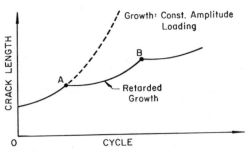

(b) Crack Growth Versus Cycle

Figure 2.42.   Crack retardation due to overloads.

Progress in fatigue fracture will now come easily as the present-day society does not support the pursuit of basic knowledge.

2.7. Appendix I. Strain energy density factor in linear elasticity

The strain energy density function $\Delta W/\Delta V$ for a linear elastic material is given by

$$\frac{\Delta W}{\Delta V} = \frac{1}{2E}(\sigma_x + \sigma_y + \sigma_z)^2 - \frac{\nu}{E}(\sigma_x\sigma_y + \sigma_y\sigma_z + \sigma_z\sigma_x) + \frac{1}{2\mu}(\tau_{xy}^2 + \tau_{xz}^2 + \tau_{zy}^2).$$

(2.105)

In linear elastic fracture mechanics, attention is given to the asymptotic stresses

$$\sigma_x = \frac{k_1}{\sqrt{2r}} \cos\frac{\theta}{2}\left(1 - \sin\frac{\theta}{2}\sin\frac{3\theta}{2}\right) + \ldots$$

$$\sigma_y = \frac{k_1}{\sqrt{2r}} \cos\frac{\theta}{2}\left(1 + \sin\frac{\theta}{2}\sin\frac{3\theta}{2}\right) + \ldots$$

$$\tau_{xy} = \frac{k_1}{\sqrt{2r}} \cos\frac{\theta}{2} \sin\frac{\theta}{2} \cos\frac{3\theta}{2} + \ldots \tag{2.106}$$

where k_1 is the crack tip stress intensity factor referred to loads applied symmetrically with respect to the crack. The polar coordinates r and θ are measured from the line tangent to the crack tip. When the loads are skew-symmetric with reference to the crack, a similar set of stresses

$$\sigma_x = -\frac{k_2}{\sqrt{2r}} \sin\frac{\theta}{2}\left(2 + \cos\frac{\theta}{2}\cos\frac{3\theta}{2}\right) + \ldots$$

$$\sigma_y = \frac{k_2}{\sqrt{2r}} \sin\frac{\theta}{2}\cos\frac{\theta}{2}\cos\frac{3\theta}{2} + \ldots \tag{2.107}$$

$$\tau_{xy} = \frac{k_2}{\sqrt{2r}} \cos\frac{\theta}{2}\left(1 - \sin\frac{\theta}{2}\sin\frac{3\theta}{2}\right) + \ldots$$

are obtained with k_2 being the stress intensity factor. A third independent set of local stresses is related to anti-plane shear loads where the top and lower crack surface are displaced in the opposite direction such that the crack edge experiences a trouser-leg motion. The resulting stresses are

$$\tau_{xz} = -\frac{k_3}{\sqrt{2r}} \sin\frac{\theta}{2} + \ldots$$

$$\tau_{yz} = \frac{k_3}{\sqrt{2r}} \cos\frac{\theta}{2} + \ldots \tag{2.108}$$

In equations (2.106) to (2.108), the crack edge coincides with the z-axis, the crack plane is normal to the y-axis and the x-axis forms an orthogonal system with y and z. Substituting equations (2.106) to (2.108) with the condition of plane strain $\sigma_z = \nu(\sigma_x + \sigma_y)$ into equation (2.105), $\Delta W/\Delta V$ takes the form S/r with

$$S = a_{11}k_1^2 + 2a_{12}k_1k_2 + a_{22}k_2^2 + a_{33}k_3^2 \tag{2.109}$$

in which the coefficients $a_{11}, a_{12}, \ldots, a_{33}$ are given by

$$16\mu a_{11} = (3 - 4\nu - \cos\theta)(1 + \cos\theta)$$

$$16\mu a_{12} = 2\sin\theta(\cos\theta - 1 + 2\nu)$$

$$16\mu a_{22} = 4(1 - \nu)(1 - \cos\theta) + (3\cos\theta - 1)(1 + \cos\theta) \tag{2.110}$$

$$16\mu a_{33} = 4.$$

The strain energy density factor is seen to depend on the Poisson's ratio ν, shear modulus μ and the polar angle θ.

2.8. Appendix II. Critical ligament length

The critical ligament length r_c for different materials may be evaluated from a knowledge of S_c in equation (2.16) and equation (2.4) which considers $\Delta W/\Delta V$ as a sum of

distortion (or yielding) occurring near the specimen surfaces, if any, occupies a small portion of the region ahead of the crack. Such a condition is known approximately as "plane strain". Obviously, the yield strength, σ_{ys}, of the material comes into play. An experimentally found relation for the thickness requirement [21] is $B \geqslant 2.5(K_{1c}/\sigma_{ys})^2$. A similar restriction holds for crack size, i.e., $a \geqslant 2.5(K_{1c}/\sigma_{ys})^2$. The dimensions of the specimens in Figures 2.43 and 2.44 are selected to satisfy these requirements in order to ensure consistent K_{1c} values. The required specimen dimensions can, in general, be determined by first estimating the anticipated K_{1c} value. A test is then performed to determine the K_1 value at fracture for the crack configuration under consideration. If both a and B are larger than $2.5(K_1/\sigma_{ys})^2$ at fracture, then the test is valid in accordance with ASTM and K_1 can be so labelled as K_{1c}.

In reality, there are many factors that can affect the measurement of K_{1c}. From the viewpoint of continuum mechanics, the three primary factors are loading, temperature and geometry while the crystallographic texture of the material at the microscopic level can also have a significant influence on the fracture toughness. Some of these factors will be discussed briefly.

Loading

One of the ways to assure that the measurement of K_{1c} is kept within the assumption of linear elastic response is to monitor the load versus displacement relation. This can be accomplished by plotting the crack opening displacement (COD) as a function of the applied load on a XY-recorder. The COD is measured by a clip gauge, equipped with electrical resistance strain gages, which is mounted into a machined receptor groove at the free end of the Chevron notch as indicated in Figure 2.45. When the

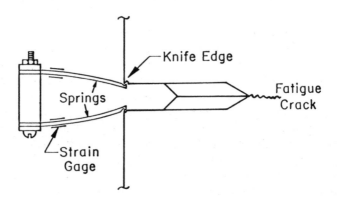

Figure 2.45   Clip gauge in Chevron notch.

specimen is loaded, the pretensioned clip gauge tends to expand with the crack opening distance. The resulting load or P versus COD curve can have different shapes depending on the operational conditions. The three commonly observed cases are shown in Figures 2.46. At first, P varies linearly with COD and the load at which unstable fracture occurs corresponds to K_{1c} (Figure 2.46(a)). Figure 2.46(b) shows that crack extension can take place at P_Q prior to total fracture. This event is audible

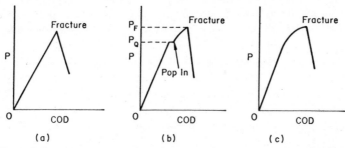

Figure 2.46. Records of P versus COD: (a) linear; (b) pop-in; (c) non-linear.

and is referred to as "pop-in". It shows up on the P-COD diagram as a plateau. Crack growth is stopped momentarily either by a drop in load or an increase in crack resistance. Load can then be further increased until fracture occurs at P_F. If excessive distortion or plasticity is present, P and COD may adopt a nonlinear relationship (Figure 2.46(c)). In such a case, the K_{1c} test method is no longer valid and the concept of stress intensity factor requires a careful reexamination. Most important of all, the cause of nonlinearity must be clearly understood before proceeding with any quantitative assessments.

Service temperature

As a rule, the fracture toughness value tends to decrease with temperature. Figure 2.47 gives a plot of K_{1c} with temperature which implies that the material toughness drops at lower temperature. This is one of the reasons why structures are more likely to fracture in cold environment. A vivid example is the Schenectady T2 tanker built during the Second World War when it fractured through the hull, with only the bottom plating holding. The ship jack-knifed. This occurred in cool weather with a recorded air temperature of 26°F and water temperature of 40°F.

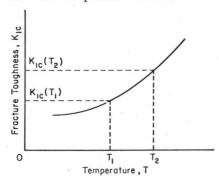

Figure 2.47. Variations of K_{1c} with temperature.

Material processing

Mechanical processing can cause structural inhomogeneities within the material which in turn leads to anisotropy of fracture toughness. In other words, K_{1c} can acquire a

practice. This question is still being debated and has been the topic of many past and present discussions. The common argument is that even though the valid K_{1c} requirements may be violated, the critical stress intensity factor quantity can still serve a useful purpose for ranking materials. The invalid data, however, should be used with caution and may not be reliable when applied to the design of engineering structures.

Table 2.2 gives the plane strain fracture toughness values of three commonly used metals: steel, aluminium and titanium. Note that those materials with high yield strength, σ_{ys}, do not necessarily correspond to high fracture toughness, K_{1c}. Depending upon the values of K_{1c} and σ_{ys}, the required thickness can vary substantially. The combination of a high fracture toughness and a low yield strength no doubt gives a large value of $(K_{1c}/\sigma_{ys})^2$ such that the ASTM recommended thickness becomes impractical. The meaning of the K_{1c} test must then be questioned.

Many real life structural members may not exhibit plane strain behaviour because of insufficient thickness and K_{1c} is no longer an adequate measure of crack resistance. Deviations from the K_{1c} criterion have been proposed and have led to much confusion in the literature. This is partly due to the restrictive nature of the K_{1c} concept that does not always conform with the growth of a natural crack in practice.

Materials with intermediate toughness and/or thinner sections usually exhibit a gradual increase of non-linearity on the load COD diagram (Figure 2.46). The non-linearity can be the cause of two factors: plastic deformation and slow crack growth prior to incipient fracture.

2.10. Appendix IV. A brief account of ductile fracture criteria

The task of defining fracture toughness in the presence of plastic deformation is difficult because of the inability to separate the energy dissipated by irreversible deformation and that used to extend the macrocrack from slow to fast growth. In principle, the energy dissipated through plastic deformation and crack extension can be measured in terms of the compliance change of the specimen for a given loading history. Such an empirical approach, however, is not useful and the result depends directly on the specimen size because the two aforementioned types of energy dissipation, i.e., deformation and fracture, cannot be separated. Their individual contribution depends on loading, specimen geometry and size. Ductile fracture is *inherently* a non-linear three-dimensional phenomenon that cannot be modelled by results obtained from two-dimensional analysis. Moreover, the stress and failure analysis must be performed simultaneously because the redistribution of stress must be accounted for each increment of loading as the material is being damaged. For more details refer to [14, 22].

Having briefly mentioned the basic features of the ductile fracture process, it is worthwhile to state the role of fracture criteria in elastic-plastic fracture mechanics:

(1) predict the entire ductile fracture process, i.e., slow and rapid crack extension;

(2) obey the laws of continuum mechanics;

(3) require minimum number of parameters based on a unique physical

assumption* that are independent loading, specimen geometry and size; and
(4) agree consistently with experiments within the regime that are proposed.

In what follows, a brief historical review of ductile fracture criterion will be given and
they will be commented upon with reference to the conditions mentioned above.

Modification of Griffith's equation

The Griffith equation being valid only for predicting the onset of catastrophic fracture
was questioned by Irwin [23] and Orowan [24] when applied to materials that
deformed plastically near the crack tips. They recommended adding a plastic surface
energy γ_p to the Griffith surface energy γ_g intended for measuring the energy required
to create a unit macro-crack surface. The argument is that since γ_p is many times
greater than γ_g, γ_g in the original Griffith equation can be simply replaced by γ_p, i.e.,

$$\sigma = \sqrt{\frac{2E(\gamma_g + \gamma_p)}{\pi a}} = \sqrt{\frac{2E}{\pi a}}\sqrt{\gamma_p} \tag{2.115}$$

which applied to a central crack specimen with a crack of length $2a$ and uniform
applied tensile stress σ.

There are two major criticisms to equation (2.115). First, γ_p addresses plasticity
or yielding which is associated with material damage at the microscopic scale. It
cannot be compared in a scalar fashion with γ_g which deals with the energy dissipated
by the macrocrack. Next, the original concept of Griffith is based on assuming that
the critical energy release rate could be used as a toughness constant. This in itself
was a major assumption for characterizing the onset of brittle fracture. Its extension
to include plasticity is not obvious and has been attempted by many past and present
investigators with no success. A detail discussion of this problem in terms of scaling
can be found in [25].

Correction on crack length

A method to include the influence of crack tip plasticity was proposed [26] by
assuming that the plastic zone is a circular region** surrounding the crack tip with a
radius

$$r_p = \frac{1}{2\pi}\left(\frac{K_{1c}}{\sigma_{ys}}\right)^2 \tag{2.116}$$

where K_{1c} is the fracture toughness and σ_{ys} the uniaxial tensile yield stress. An
effective crack length

$$a_{eff} = a + r_p \tag{2.117}$$

is defined with r_p being a correction for the effect of plasticity. This approach is still

*The arbitrary assumption that two separate failure criteria in a single physical problem can lead to
inconsistent results..
**This is obviously an oversimplified assumption for Mode I crack tip stress field where the plastic
zones are off to the side of the crack. In fact, there is little or no yield of the material directly
ahead of the crack under plane strain.

inconceivable that the same investigators who admit numerical inaccuracy as the cause for path dependence claim accuracy when reporting calculated values of crack tip opening displacements that are, in fact, in the same region where J values are admitted to be poor.

There are too many fundamentally unresolved difficulties concerning the association of J with ductile fracture. In the case of plasticity, it certainly has no relationship with the macrocrack energy release rate. Even for the elastic case, it is limited to two dimensions and cracks that must propagate in a self-similar manner. The arguments for supporting J as a ductile fracture criterion being self-contradictory and/or inconclusive are in themselves evidence of the inability of J to characterize ductile fracture.

Resistance curve

Attempts have been made to extend the J concept to include slow stable crack growth by assuming that slow growth occurs as

$$\frac{\Delta J_r}{\Delta a} = \text{const.} \tag{2.120}$$

where J_r is the measured value of J. Equation (2.120) is obtained from a known resistance curve representing a plot of J_r versus a (crack length) [33, 34]. It is also proposed that stable growth continues until J_r reaches J_c for the global instability of the specimen. This approach requires that the resistance curve be linear for the range of stable crack growth which is not demonstrated epxerimentally [35, 36]. The suggestion of applying some critical value of J_r to predict crack growth initiation is also not useful because its strong dependence on specimen geometry [36, 37].

Crack opening displacement

The critical crack opening displacement (COD) has been proposed [38] as a fracture criterion. The opening is normally referred to a region near the crack tip. For a Mode I crack in a linear elastic material, the vertical displacement of the crack surfaces near the tip is

$$v_c = \frac{2\sqrt{2r}(1-v^2)K_{1c}}{\sqrt{\pi E}} \tag{2.121}$$

where r is the radial distance measured from the crack tip. Since equation (2.121) drastically underestimates experimental measurement, the concept of a fictitious crack elongated by the length of a fictitious yield zone ahead of the crack is used. This is to replace r in equation (2.121) by r_p in equation (2.116) which leads to

$$\delta_c = \text{COD} = \frac{4(1-v^2)K_{1c}^2}{\pi E \sigma_{ys}} \tag{2.122}$$

where $\delta_c = 2v_c$. Early experimental evidence [39] suggested that COD gives a reasonable prediction of global instability if the amount of yielding near the crack tip is sufficiently small. Difficulty in the measurements of COD made the approach inconvenient and led to much confusion and contradiction in the literature. Improved

98

experimental techniques have allowed more accurate and consistent measurement of COD.

The shortcomings of the COD criterion are similar to those criteria mentioned earlier. It does not account for specimen thickness effect and slow crack growth which is an inherent feature of ductile fracture. Another serious limitation is that inability of the criterion to transfer measured data to the design of structural components.

In order to account for slow growth, the concept of COD resistance curve [40] has been suggested:

$$\frac{\Delta\delta}{\Delta a} = \text{known quality.} \tag{2.123}$$

Initiation of growth is assumed to occur when δ reaches a critical value until it coincides with δ_c for global instability. Again, the dependency of the initiation value and δ_c on specimen thickness and geometry makes the criterion suggested by equation (2.123) inadequate.

Attempts have been made to correlate J and COD values both experimentally and analytically [41]. The general consensus is that critical COD and J values are linearly related if yielding is kept small. They do not have a direct correlation as yielding increases. In general, these criteria seem to be applicable only when the crack tip state stress is perturbed only slightly from linear elasticity.

Strain energy density function

This criterion has already been discussed in detail in the present communication. Hence, it suffices only to emphasize its ability to characterize the complete ductile fracture process in a consistent fashion from initiation to onset of rapid fracture by the single relation $(\Delta W/\Delta V)_c = S_c/r_c$. The critical ligament length r_c designates the transition from slow growth initiated by reaching a value $(\Delta W/\Delta V)_c$ to rapid fracture governed by S_c.

In conclusion, the usefulness of a fracture criterion should be judged by the simultaneous satisfaction of the four conditions mentioned earlier.

References

[1] Sih, G.C., "The Role of Fracture Mechanics in Design Technology", *Journal of Engineering for Industry,* 98, p. 1243 (1976).

[2] Sih, G.C. (ed.), *Mechanics of Fracture,* Volume I to VII, Sijhoff and Noordhoff International Publisher, The Netherlands (1973 to 1981).

[3] "The Determination of Absorbed Specific Fracture Energy", *Hungarian Standard MSz 4929-76* (English Translation), Institute of Fracture and Solid Mechanics Publication, Lehigh University (1978).

[4] Gillemot, L.F., "Criterion of Crack Initiation and Spreading", *Journal of Engineering Fracture Mechanics,* 8, p. 239 (1976).

[5] Czoboly, E., Havas, I. and Gillemot, F., "The Absorbed Specific Energy Till Fracture as a Measure of the Toughness of Metals", *Proc. of Int. Symp. on Absorbed Specific Energy and/or Strain Energy Density Criterion,* edited by G.C. Sih, E. Czoboly and F. Gillemot, Sijthoff and Noordhoff International Publisher, The Netherlands, p. 107 (1981).

[6] Sih, G.C., "Three-Dimensional Stress State in a Cracked Plate", *International Journal of Fracture Mechanics*, (1), p. 39 (1971).

[7] Sih, G.C., "The Analytical Aspects of Macrofracture Mechanics", *Proc. of Int. Conf. on Analytical and Experimental Fracture Mechanics*, edited by G.C. Sih and M. Mirabile, Sijthoff and Noordhoff International Publisher, The Netherlands, p. 3 (1981).

[8] Sih, G.C. and Kipp, M.E., "Fracture Under Complex Stress – The Angled Crack Problem", Discussion, *International Journal of Fracture* (2), p. 261 (1974).

[9] Sih, G.C., Wei, R.P. and Erdogan, F. (eds.), *Linear Fracture Mechanics*, Envo Publishing Co., Inc., Lehigh Valley, Pennsylvania (1976).

[10] Sih, G.C., Villarreal, G. and Hartranft, R.J., "Photoelastic Investigation of a Thick Plate with a Transverse Crack", *Journal of Applied Mechanics*, 42, p. 9 (1975).

[11] Erdogan, F. and Sih, G.C., "On the Crack Extension in Plates Under Plane Loading and Transverse Shear", *Journal of Basic Engineering*, 85, p. 519 (1963).

[12] Sih, G.C. *Handbook of Stree-Intensity Factors*, Institute of Fracture and Solid Mechanics, Lehigh University (1973).

[13] Sih, G.C. and Macdonald, B., "Fracture Mechanics Applied to Engineering Problems – Strain Energy Density Fracture Criterion", *Journal of Engineering Fracture Mechanics*, 6, p. 361 (1974).

[14] Sih, G.C., "Mechanics of Crack Growth: Geometrical Size Effect in Fracture", *Proc. of Int. Conf. on Fracture Mechanics in Engineering Application*, edited by G.C. Sih and S. R. Valluri, Sijthoff and Noordhoff International Publisher, The Netherlands, p. 3 (1979).

[15] Paris, P.C. and Erdogan, F., "A Critical Analysis of Crack Propagation Laws", *Journal of Basic Engineering*, 88, p. 528 (1963).

[16] Sih, G.C., "Experimental Fracture Mechanics: Strain Energy Density Criterion", *Mechanics of Fracture*, Vol. 7, Edited by G.C. Sih, Sijthoff and Noordhoff International Publisher, The Netherlands, p. XVII (1981).

[17] Dill, H.D. and Saff, C.R., "Environment-Load Interaction Effects on Crack Growth", *AFFDL-TR-78-137* (1978).

[18] Sih, G.C. and Barthelemy, B., "Mixed Mode Fatigue Crack Growth Predictions", *International Journal of Engineering Fracture Mechanics*, 13, p. 439 (1980).

[19] Peterson, R.E., "Stress Concentration Factors; Charts and Relations Useful in Making Strength Calculations for Machine Parts and Structural Elements", John Wiley and Sons, New York, New York (1974).

[20] Pustejovsky, M.A., "Fatigue Crack Propagation in Titanium Under General In-Plane Loading – I: Experiments", *Engineering Fracture Mechanics*, 11, p. 9–15 (1979).

[21] Brown, W.F., Jr. and Srawley, J.E., (eds.), "Plane Strain Crack Toughness Testing of High Strength Metallic Materials", *ASTM Special Technical Publication No. 410* (1966).

[22] Sih, G.C., "Fracture Toughness Concepts", *Properties Related to Fracture Toughness STP 605*, American Society of Testing and Materials, Philadelphia (1976).

[23] Irwin, G.R., *Fracture of Metals*, American Society of Metals, Cleveland, Ohio (1948).

[24] Orowan, E., "Energy Criterion of Fracture, Welding Criterion of Fracture, *Welding Research Supplement*, p. 157 (1955).

[25] Sih, G.C., "Some Basic Problems in Fracture Mechanics and New Concepts", *Journal of Engineering Fracture Mechanics*, 5, p. 365 (1973).

[26] Irwin, G.R., "Structural Mechanics", *Proc. First Symp. Naval Struct. Mech.*, Pergamon Press, New York, p. 557 (1958).

[27] McClintock, F.A., "The Plasticity Aspects of Fracture", *Fracture*, Vol. III, edited by H. Liebowitz, Academic Press Book Co., New York, p. 47 (1971).

[28] Rice, J.R., "A Path Independent Integral and the Approximate Analysis of Strain Concentration by Notches and Cracks", *Journal of Applied Mechanics*, 35, p. 379 (1968).

[29] Eshelby, J.D., "The Calculation of Energy Release Rates", *Prospects of Fracture Mechanics*, edited by G.C. Sih, H.C. van Elst and D. Broek, Noordhoff International Publisher, The Netherlands, p. 69 (1974).

[30] Turner, C.E., *Post Yield Fracture*, edited by D.G.H. Latzgo, Appl. Sci. Pub. Lts., p. 23 (1979).

[31] Begley, J.A. and Landes, J.E., "The J Integral as a Fracture Criterion", *Fracture Toughness*, ASTM STP 514, p. 1 (1972).

100

[32] Wilson, W.K. and Osias, J.R., "A Comparison of Finite Element Solutions for an Elastic-Plastic Crack Problem", *International Journal of Fracture,* 14, p. R95 (1978).

[33] Bucci, R.J., Paris, P.C., Landes, J.D. and Rice, J.R., "J Integral Estimation Procedure", *Fracture Toughness,* ASTM STP 514, p. 40 (1971).

[34] Turner, C.E., "Description of Stable and Unstable Crack Growth in the Elastic-Plastic Regime in Terms of J_r Resistance", *Fracture Mechanics,* ASTM STP 677, p. 614 (1979).

[35] Garwood, S. J. and Turner, C.E., "The Use of the J-Integral to Measure the Resistance of Mild Steel to Slow Stable Crack Growth", *Fracture,* ICF-4, Vol. 2, p. 279 (1977).

[36] Willoughby, A.A., Pratt, P.L. and Turner, C.E., "The Effect of Specimen Orientation on the R-Curve", *International Journal of Fracture,* 14, p. R249 (1978).

[37] Wilhem, D.P., "J-Integral Approach to Crack Resistance for Aluminium Steel and Titanium Alloys", *Journal of Mechanical Engineering Science and Technology* (Trans. ASME, Vol. 99), p. 97 (1977).

[38] Wells, A.A., "Unstable Crack Propagation in Metals: Cleavage and Fast Fracture", *Proc. of the Crack Propagation Symposium,* Cranfield, p. 210 (1961).

[39] Wells, A.A., "Application of Fracture Mechanics At and Beyond General Yielding", *British Welding Journal,* 10, p. 563 (1963).

[40] Harrison, J.D., *Metal Construction,* Part 1, p. 415 (1980), Part 2, p. 524 (1980) and Part 3, p. 600 (1980).

[41] Shih, C.F., DeLorenzi, H.G. and Andrews, W.R., "Studies on Crack Initiation and Stable Crack Growth", *Elastic-Plastic Fracture Mechanics,* ASTM STP 668, p. 65 (1979).

Failure mechanics: damage evaluation of structural components

3.1. Introduction

The main content of this book deals with the science of fracture mechanics. Methods are presented — many of them highly sophisticated — for analyzing the growth of subcritical cracks and for predicting failure loads via either linear or ductile fracture theory. This chapter deals with failure mechanics, viz: how to apply sciences like fracture mechanics to resolve structural failures which may occur in fleets in service.

Failure mechanics today draws heavily upon fracture mechanics because system designers over the past two decades have demanded increasing performance from structures and materials. Damage modes other than sharp cracks still control structural failures in many cases, and the failure analyst must beware of a too hasty focus on crack growth or any other particular damage mechanism. However, applied failure mechanics is a common ground beyond this point.

When a fleet in service experiences a significant structural failure, safety and operational considerations may conflict. While safety considerations may suggest that the fleet be immediately removed from service, the failure might have been an isolated event, and maintaining the ability to perform a vital strategic mission or to continue an economically important public-sector service may require continued operation while the failure is being studied. The failure analyst thus faces two challenges: to define immediate actions which will reduce the operational risk to a tolerable level and to find a permanent solution as quickly as possible.

The body of this chapter elaborates the foregoing thoughts in the light of the author's experience with failure investigations. The following section presents in its entirety a short case which illustrates both the importance of focusing on the critical damage mode and the use of short cuts for quick resolution. The remainder of the paper covers service load description and life prediction. Approximate methods are suitable for these tasks because failure mechanics — unlike theoretical and experimental fracture mechanics — rarely requires precise answers. Some short cut methods which the author has found useful are presented, and their limits of applicability are noted.

3.2. Failure of a railroad passenger car wheel

Figure 3.1 shows a wheel fracture which was reported to have caused the derailment of a railroad passenger coach in early 1978, resulting in several injuries. Close examination of the fracture surface reveals an origin at the outer edge of the wheel flange (Figure 3.2). A corner crack about one half inch (13 mm) in radius with no shear lip is located here, indicating that fatigue crack growth had occurred over this small region. The wheel is a monolithic low-carbon steel casting, but the tread has been hardened by intense plastic flow. The nominal material and fabrication are standard for passenger coach wheels, and the tread plastic flow is expected from normal wheel/rail contact pressures in service.

However, the corner crack failure is an uncommon event. In the passenger coach fleet in question, the fractured wheel is one of about 300 wheels manufactured by a particular supplier; 300 wheels represent about 5—10% of the wheels in the fleet.

One's natural first reaction to the foregoing situation is to perform a fracture mechanics analysis to determine the rate of growth of the corner crack. Although complicated by the wheel geometry and local plastic flow, such an analysis would undoubtedly confirm the suspicion that there is little time available to detect the crack before it reaches critical size, given the level of inspection technology which can reasonably be supported in a railroad maintenance environment.

Safety requirements would thus dictate that the affected wheels be removed from service immediately, until the 300 fracture-sensitive wheels could be replaced. There was great pressure to do so immediately after the 1978 accident, even though the action would have entailed severe cost and service reliability penalties. The action would have affected 20—40% of the fleet vehicles because axles with wheels from different suppliers are routinely mixed on individual coaches.

However, further investigation of the failure led to a quite different conclusion. Examination of the opposite wheel on the derailed axle revealed a through radial crack (Figure 3.3, arrow A) which had released the wheel/axle interference fit and

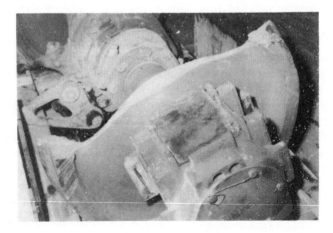

Figure 3.1.   Fractured wheel.

Figure 3.2.   Detail showing fatigue crack growth region.

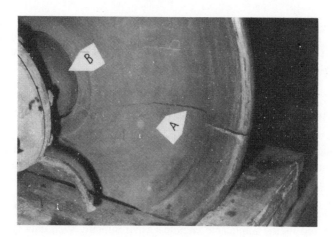

Figure 3.3.   Opposite wheel. (A) fully developed radial crack; (B) partly exposed seat shows that wheel has shifted inboard on axle).

allowed this wheel to shift inboard (arrow B). Shifting of the loose wheel by normal wheel/rail lateral guiding forces was then deemed to be the likely cause of the accident. This cause was quickly confirmed by reassembling the fractured wheel (Figure 3.4) to show that two pairs of post-derailment wheel/rail impact marks lined up in a way that can have happened only if the corner crack fracture had occurred after the derailment.

The real culprit in this case is a well understood phenomenon. Radial cracking tends to appear in railroad car wheels which have been severely overheated by abnormal tread brake application. Signs of the overheating can often be detected after the event by observing a wheel discoloration which differs from the normal rust coating, but in any case many years of service experience have shown that these radial cracks grow slowly and can be reliably detected by hammer-tap inspection well before a wheel

Figure 3.4.   Reassembly of fractured wheel (two sets of marks on rim identify impacts which caused the final fracture).

press-fit would be relieved. In the light of these additional facts, the failure was resolved by increasing supervision of normal inspections to assure that subcritical radial cracks would be reliably detected. To date this fleet has not experienced any other derailments caused by cracked wheels.

3.3. Describing the load environment

The first quantitative task in a failure investigation is to describe the structure's service load environment. Thermal and chemical components of the environment often have significant influence on structural damage but can usually be treated as steady-state effects. For example, structures fabricated from corrosion-sensitive high-strength alloys are dealt with by estimating damage rates based on materials data taken from specimens tested in the applicable chemical environment. Such data is either readily available [1, 2] or can easily be gathered from standard laboratory tests.

Considerable effort must be devoted, however, to describing the structure's load environment, which is usually the dominant factor in failure, is generally rich in its variety of occurrence patterns, is always unique to the type of structure involved, and is only partly projectable from experience with similar structures. The objective of characterizing loads for a failure investigation is to provide a reasonable summary for assessment of damage rate tradeoffs with respect to changes in usage or modifications to reduce design-stress levels. Time and resource restrictions associated with failure investigations often demand the simplest load summary compatible with the primary objective of resolving the problem.

The simplest summary of the repeated-load type of environment associated with mechanical failures involves three parameters: frequency of occurrence, minimum load, and succeeding maximum load. The load values can be described in the equivalent alternative forms: mean and amplitude; mean and range; or range and minimum/maximum ratio. One usually has to deal with a mean value based on static calculations and with statistical estimates for expected occurrence frequency and deviation of dynamic load from the mean to account for the inevitably random nature of service loads. This type of simplified summary is adequate for many but not all situations. After some important exceptions have been noted, the discussion will focus on estimation short cuts and data interpretation.

Spatially distributed loads

Most structural service loads are continuously or discretely distributed and are usually composed of at least two partly coherent or uncorrelated sources. For example, the bus undercarriage shown in Figure 3.5 is subject to dynamic inputs from road surface irregularities at the four road/tire contact points. The random elements in the road irregularities make the inputs only partly coherent [3]. Hence, if the problem at hand involves fatigue crack growth in one of the undercarriage beams, the input summary must be expanded to include phase relations, and one also needs the dynamic influence functions for the critically stressed location.

It is possible in principle to approach the investigation as suggested above, i.e., by synthesizing a road input summary from measurements and building a dynamic model of the bus. This approach is attractive in design because the investment in data synthesis and dynamic modelling is amply repaid by the ability to perform a thorough detail fatigue analysis at hundreds of stress locations.

In a failure investigation, however, urgency forbids the investment and in any case the fleet failure records have already identified the most critically stressed locations. One should, therefore, let the structure take care of the phase relations by relying on strain gauge data at the known critical points.

This alternate approach is practical, but inherently risky in that the fleet failure information is incomplete unless most of the service life has already elapsed. One should also employ simplified static and dynamic analyses, therefore, to extend the strain gauge data coverage and reduce the risk of missing a location that may become critical in the future.

Load sequence effects

Some damage mechanisms have rates that depend on the load-time history as well as the current load cycle. The important mechanisms in this category include low-cycle fatigue crack nucleation, high-cycle fatigue (when analysis is based on S-N curves), fatigue crack growth in retardable alloys subjected to certain types of isolated-overstress load histories, and creep-fatigue interactions. Although it is not a direct cause of failure, one should also include elastic-plastic shakedown because this mechanism can create large residual stresses which may later contribute to fatigue crack nucleation or growth. The effect of rolling contact on redistribution of residual stresses in the heads of railroad rails is an example of this type of shakedown [4].

Little additional effort is required to extend a three-parameter load summary to cover sequence-dependent situations if one is willing to accept representative sequences based on judgements made in the light of operating or mission profiles. Such judgements are a routine practice in fatigue crack growth assessment, for example;

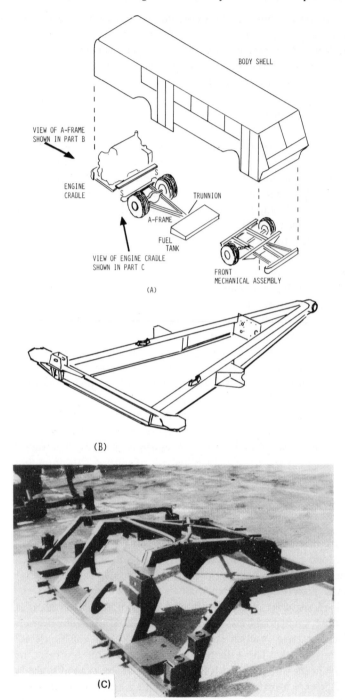

Figure 3.5. A light-weight bus fleet in which the undercarriage developed early fatigue cracks. (A) general schematic of body, suspension, and undercarriage components; (B) configuration of 'A-frame' before modification; (C) photograph of modified engine cradle.

the practice is reasonable because the sequence details tend to repeat many times in a service life.

On the other hand, using a sequence-dependent damage model to estimate safe lifetimes entails a considerable investment in book-keeping and computing. The investment is not justified for a failure investigation when the damage mechanism does not depend on sequence or when the dependence has little influence on life. Mechanisms such as fretting corrosion and abrasive wear do not depend on sequence. High-cycle fatigue crack nucleation life can be estimated with acceptable accuracy if the calculations are based on strain-range-N curves [5], and even S-N curve calculations are useful for order-of-magnitude estimates. As will be seen later, most fatigue crack growth situations involve material and load combinations for which acceptably accurate life estimates can be made while neglecting sequence effects.

Statistical consistency

The equivalence of alternative peak descriptors was mentioned earlier, namely:

$$S_M = (S_{max} + S_{min})/2 \qquad (3.1)$$

$$S_A = (S_{max} - S_{min})/2 \qquad (3.2)$$

and

$$\Delta S = 2 S_A \qquad (3.3)$$

$$R = S_{min}/S_{max} \qquad (3.4)$$

where stress, S, is used here for illustration, and where S_M, S_A, ΔS, and R are respectively the mean, amplitude, range, and ratio parameters. One must be careful, however, when attempting to use these relations to derive equivalent statistical values.

For example, a report of laboratory experiments in which the fatigue crack growth rates of bridge steels were correlated with the RMS value of ΔK in random-load tests [6] has led to a proposed life prediction approach based on RMS stress parameters defined in the following manner [7]:

$$S_{RMS} = S_{max, RMS} - S_{min, RMS} \qquad (3.5)$$

$$R_{RMS} = S_{min, RMS}/S_{max, RMS} \qquad (3.6)$$

where $S_{min, RMS}$ and $S_{max, RMS}$ are estimated from the raw variance of peak counts:

$$S_{P, RMS} = \sqrt{\frac{1}{N} \sum_{i=1}^{N} (S_{P, i})^2} ; \qquad P = \text{min or max.} \qquad (3.7)$$

A simple numerical application will serve to illustrate the fallacy in applying the parameter transformation after the statistics have been estimated, as is done in the foregoing procedures. Consider a crack growth analysis of a typical aircraft aluminum alloy with a $(\Delta K)^4$ growth rate behavior and suppose that the service loads can be treated as a zero-mean narrow-band Gaussian process. Then the distribution of stress ranges can be associated with the peak distribution, which is a Rayleigh process, i.e., the probability density function for the ranges is given by:

$$f(\Delta S) = f(S_{max}) = (S_{max}/\sigma^2) \exp(-(S_{max})^2/2\sigma^2) \tag{3.8}$$

where $\Delta S = 2S_{max}$, and hence:

$$\overline{(S_{max})^2} = 0.429\sigma^2 \qquad \Delta S_{RMS} \cong 1.31\sigma. \tag{3.9}$$

However, the life prediction should be based on a mathematically proper expectation of the crack growth rate [8], which requires the root mean fourth value in this case. For the Rayleigh distribution, this leads to:

$$\overline{(S_{max})^4} = 3.26\sigma^4 \qquad [\overline{(\Delta S)^4}]^{1/4} \cong 1.60\sigma. \tag{3.10}$$

While the difference between the RMS and root mean fourth values is not great, the predicted crack growth rates and lifetimes will differ by the fourth power of the ratio of these parameters, i.e., by a factor of 2.2.

Counting

Service load summaries are often derived by applying counting methods to load-time histories. Simple methods such as counting peak magnitudes, mean-crossing peaks, and level crossings have been 'hard-wired' in automatic counters that can be left unattended for long times. The large-sample capabilities of such devices are useful, but the simple methods can miss significant information [9] and transforming the summary statistics into cyclic load parameters can compound the error.

Counting errors can be minimized by making judicious adjustments to some of the other simple methods if one has prior information about the general characteristics of the service loading and some idea about the relative significance of its components with respect to the relevant damage mechanism. For example, the racetrack method (level or peak counting with all but the top 10–20 percentile values censored) can gather useful information for structures with design-stress levels well below an endurance strength and service loading containing isolated overloads [10]. The level-discrimination method* (level exceedance counted only when the load drops below an associated lower level) can be adjusted to eliminate high-frequency (HF), low-amplitude data from service loading that is dominated by LF/high-amplitude content.

The U.S. Air Force employs level-discrimination counting accelerometers located near the airplane center of mass in its Individual Airplane Tracking Program (IATP). These devices accumulate counts at about six levels on a set of mechanical dials which are read and reset monthly on the flight line. The IATP is an anticipatory program; the accelerometer counts are processed to enable the Air Force Logistics Command to track each airplane in terms of an estimated damage accumulation parameter and to fine-tune depot maintenance schedules. Acceleration level-discrimination counts are particularly useful for fighters because the wing stresses in these airplanes are dominated by VLF maneuver loads that are highly coherent with accelerations of the airplane center of mass.

---

*Also sometimes called the fatigue-meter method.

The utility of level-discrimination counting can be extended to situations where the LF loading component is less clearly dominant than in the fighter airplane case, if one is willing to record and post-process the data. The first step is to digitize the data at a sampling rate at least four times the highest significant load frequency. The analysis is then performed on a digital computer with level-discrimination software. Since digital computers have large data capacities, this approach also allows one to monitor a large number of levels simultaneously for better resolution, to count min/max pairs, and even to preserve sequence information. This approach has been used to analyze railroad passenger coach truck-bolster loads with the discrimination function adjusted to eliminate amplitude excursions that were small in proportion to the individual major cycles [11].

The general characteristics of many kinds of service loading are known well enough to justify simple counting methods and other short cuts. (See Figure 3.6 for several examples of service loads.) However, one does not always have such confident knowledge, and the loading may turn out to have modulated or broad-band characteristics that make such short cuts difficult to apply. One must then rely on complex methods such as rainflow or range-pair counting to produce summaries that have the proper meaning for damage analysis [9]. The complex methods require data buffering over a significant time-window, however, e.g., on the order of one mission to pick up the maneuver-load range-pairs seen by a fighter airplane. Hence, the complex methods are less suitable for unattended counting than the simple methods because of the risk of data loss during transient power dropouts. However, the complex methods are quite useful and easy to apply as digital or analog post-processors.

Spectral analysis

Counting is complemented by spectral analysis, which allows the study of phase and cause-effect relations between load sources and stress locations. Unlike counts, spectral information can be used to infer the effects on stress when part of the operating environment changes or when the investigation focus must be shifted to an uninstrumented stress location.

For example, axles and driveshafts are generally subject to bending and torsional stresses that have partly coherent sources. Strain gauges can be used to individually measure the bending stress, S_B, and the shear stress due to torque, S_Q, but the measurements must include phase and cause-effect information if one wants to predict (say) the change in S_B if drive-motor torque is increased beyond the test range. The phase relation is also essential if one wants to prepare a summary of a fatigue-related parameter such as octahedral stress:

$$S_{oct} = \sqrt{(S_B)^2 + 3(S_Q)^2}. \tag{3.11}$$

The bus undercarriage investigation mentioned earlier [3] provides another illustration. The only service data available in that case were strain gauge level-discrimination counts. Data analysis revealed that the engine cradle framing (Figure 3.5) was subject to bending moments in two planes caused by engine roll and pitch, as well as by carbody vertical motions. The effects of the carbody source could be extrapolated

LEGEND

SIGNAL _____ TIME   AUTO-SPECTRAL DENSITY _____ FREQUENCY

(A) NARROW-BAND PROCESS
(B) TRANSPORT AIRPLANE LOWER WING SKIN STRESS
(C) RAILROAD FREIGHT CAR TRUCK-BOLSTER SPRING LOAD
(D) STRESS IN A RAIL HEAD (DIURNAL CYCLE FREQUENCY EXAGGERATED IN TIME PLOT FOR CLARITY)
(E) HIGH-FREQUENCY PROCESS MAGNITUDE-MODULATED BY A LOW FREQUENCY
(F) LF/HF PROCESSES PRODUCING A MODULATED COMBINATION
(G) BROADBAND PROCESS
(H) FIGHTER AIRPLANE LOWER WING SKIN STRESS

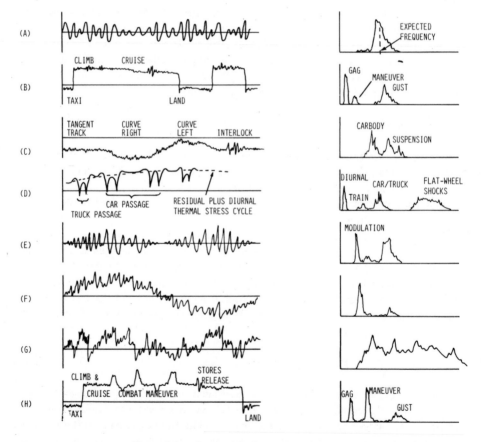

Figure 3.6.   Random loading characteristics.

from a strain gauge to neighboring points by means of static analysis. It was impossible to extrapolate the engine-motion effects, however, because the roll and pitch contributions could not be separated to permit the use of unit-load analyses. Conservative estimates had to be made to bound the damage analysis, which fortunately predicted ample life margins. Had the margins been on the borderline of service life, however, spectral analysis would have been essential to complete the investigation.

The spectral analysis method is based on applying the Fast Fourier Transform technique to convert time-history samples to the frequency domain. Autospectral density (load^2/Hz) is the transformed counterpart to load versus time for an individual signal, and contains both occurrence frequency and RMS information. Phase and cause-effect relations are produced by comparing two signals in terms of cross-spectral density, transfer function, and coherence function. Reference [12] contains a comprehensive theoretical and practical treatment of spectral analysis techniques and data interpretation.

Development of advanced spectrum analyzers over the past twenty years has greatly reduced the labor burden of such analyses. Modern dual-channel analyzers have internal wideband A/D converters and a variety of push-button functions that make the equipment compatible with standard FM analog tape recorders and easy to use interactively with mission-length data samples. The equipment utilizes buffered data and is thus sensitive to power dropouts. A spectrum analyzer is most convenient for post processing, but can also be useful for 'quick-look' preliminary data analysis if a test/engineering crew is onboard.

Modern analyzers usually have some counting abilities as well as the spectral functions. However, the counting is commonly limited to the standard probability density and cumulative distribution functions. These functions count the time a level is exceeded rather than the number of discrete events one obtains from the methods discussed in the preceding section. The probability functions can be used to infer summary load parameters, but occurrence-rate information is lost and must be recovered independently.

The following example illustrates the flexibility offered by a spectrum analyzer. In a case involving a fleet of 500 railroad passenger coaches, the disc-brake attachments were failing at about 10% of the vehicle service life by a fretting corrosion mechanism (Figure 3.7). During the investigation there arose a question of whether the failures resulted from a design that was inadequate for the nominal service environment or were caused by unexpectedly hard usage. Those in favor of the hard-usage hypothesis pointed to the fact that the fleet wheels had a larger than normal population of tread defects (Figure 3.8). The essence of the argument was: (1) the defects persisted because of poor operator maintenance on the non-tread-braked wheels (in tread-braked applications the brake pads continuously clean up wheel defects); and (2) the tread defects were the dominant contributor to axle dynamic loading.

A spectral analysis ultimately answered the question [13]. An additional test was run with accelerometers placed on the bearing nearest to a wheel which had been specially selected for the test. This wheel contained a large tread spall which the test crew could clearly hear as it pounded the rail on each revolution. Figure 3.9 illustrates the test configuration.

In the post-test analysis the spectrum analyzer was set to passively monitor the

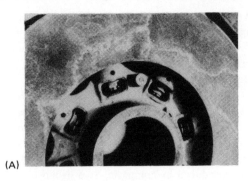

(A)

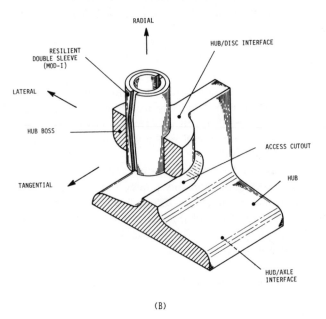

RADIAL

RESILIENT
DOUBLE SLEEVE
(MOD-I)

HUB/DISC INTERFACE

LATERAL

HUB BOSS

ACCESS CUTOUT

HUB

TANGENTIAL

HUB/AXLE
INTERFACE

(B)

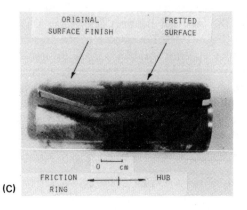

ORIGINAL
SURFACE FINISH

FRETTED
SURFACE

0   cm

FRICTION
RING

HUB

(C)

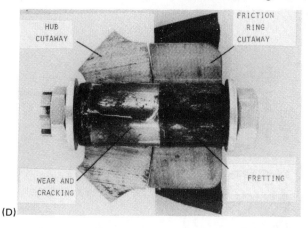

Figure 3.7. Early fretting failure of a railroad disc brake. (A) arrangement of hub to friction ring attachments; (B) cutaway view of resilient-sleeve interference fit; (C) a sleeve in the initial stage of fretting corrosion; (D) a sleeve which has worn beyond the fit tolerance and is close to fracture.

Figure 3.8. Wheel condition after service without tread brakes (a typical colony of spalls appears at the top of the photograph).

signal until a large spike occurred. Data from the spike onward was then captured for a block consisting of two wheel revolutions, and the procedure was automatically repeated for 16 samples. Finally, the samples were averaged to eliminate random data, leaving only the response to spall impacts at one-per-revolution and harmonic

115

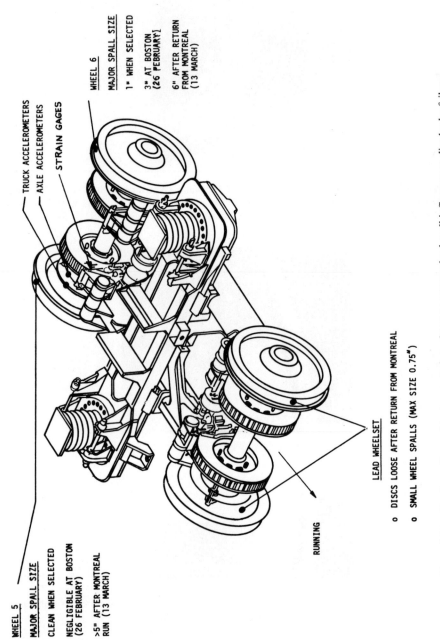

TRUCK ACCELEROMETERS
AXLE ACCELEROMETERS
STRAIN GAGES

WHEEL 6
MAJOR SPALL SIZE
1" WHEN SELECTED
3" AT BOSTON
(26 FEBRUARY)
6" AFTER RETURN
FROM MONTREAL
(13 MARCH)

WHEEL 5
MAJOR SPALL SIZE
CLEAN WHEN SELECTED
NEGLIGIBLE AT BOSTON
(26 FEBRUARY)
>5" AFTER MONTREAL
RUN (13 MARCH)

LEAD WHEELSET
o DISCS LOOSE AFTER RETURN FROM MONTREAL
o SMALL WHEEL SPALLS (MAX SIZE 0.75")

RUNNING

Figure 3.9.   Configuration and instrumentation of test to assess wheel-spall influence on disc brake failures.

frequencies. Figure 3.10 is a typical typical 16-sample time average showing the periodicity of the filtered response.* The analysis proved that the filtered RMS acceleration was much lower than the unfiltered RMS, and the failure was attributed to inadequate design.**

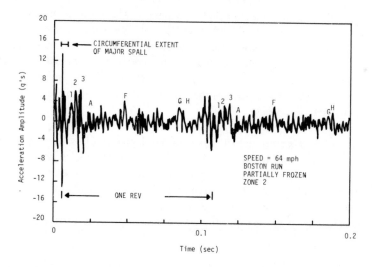

Figure 3.10. Bearing acceleration-time history enhanced to highlight contributions related to wheel revolution.

3.4. Interpreting service load data

The atmosphere of urgency in a failure investigation often means that answers must be based on incomplete information. Typically one must reconstruct a load summary from spectral or unattended counter data. Even if the operative damage mechanism does not require sequence information, one must still reconstruct a reasonable picture of the min/max pairs or their equivalents to predict safe life.

It is usually possible to make a reasonable reconstruction by taking advantage of general knowledge about structural dynamic behavior and the characteristics of the service loading. The zero-mean narrow-band Gaussian process (the simplest form of random loading) provides a good conceptual framework for reconstruction, and is taken as a starting point in the next section. The sequel deals with reconstructions for practical cases and shows that the narrow-band Gaussian process or a variation on its theme is often sufficient.

---

*The events marked 1, 2, . . . , *G*, *H* were later correlated with the locations of smaller defects at other points on the circumference of the tread.
**The design was later modified to a taper-lock attachment, which appears to have eliminated the fretting problem.

The narrow-band Gaussian process

The time- and frequency-domain characteristics of the zero-mean narrow-band Gaussian process were illustrated in Figure 3.6. The simplicity of this process lies in the fact that all of its characteristics are embodied in two parameters: the variance σ^2 (or RMS value σ) and the expected frequency $\bar{f}$. The variance appears as the area under the autospectral density function and as the scale parameter in the probability density function, the level exceedance curve, and the peak occurrence density curve. The expected frequency lies approximately at the midpoint of the autospectral density function and also scales the ordinates of the level exceedance and peak occurrence density curves. Table 3.1 summarizes the definitions and mathematical expressions for zero-mean Gaussian processes.

Of central importance to the job of damage analysis is the tendency for neighbouring peaks to be strongly correlated in a narrow-band process. When combined with the symmetry of positive and negative peak distributions in a Gaussian process, this means that the positive peak occurrence density curve also defines the occurrence rates of cycle amplitudes and ranges. Thus, preparing a service load summary from either an autospectral density plot or from properly acquired counting information is a routine task if one is willing to assume that the data sample represents a narrow-band Gaussian process.

Short cuts with narrow-band loading

Random-load sources such as guideway irregularities or air turbulence are generally broad-band non-Gaussian processes. In spite of this, one can often assume that the loading itself has a narrow bandwidth because the vehicle structure acts as a filter and the loads are measured at a dynamically isolated point.

For example, components in the sprung mass of a railroad passenger-coach truck are isolated from the rails by primary suspensions which typically attenuate at 5 to 10 Hz. The axles of buses and trucks are similarly isolated from the roadway by tire flexibilities between 10 and 30 Hz. Since these isolators are moderately damped (typical damping factors range from 0.05 to 0.2), the sprung structure sees only filtered narrow-band loads.

The response of a transport airplane wing to turbulence is another example of narrow-band loading. In this case the wing bending/torsion natural modes act as the isolators and unsteady aerodynamics provides the damping. Gust loads typically account for about half of the fatigue damage in transport wings. The other half is caused primarily by the ground-air-ground cycle (see Figure 3.6), which can be treated deterministically.

The main decision one must make about reconstructing service loads for isolated structure is whether the loading is close enough to a Gaussian process to justify approximating the cycle distribution. A level exceedance curve without apparent inflection points when plotted on semilogarithmic paper is a good indicator.

Figure 3.11 shows level strain counts taken on a bus undercarriage beam during simulated revenue operations on a city street. The data seems to be Gaussian with a positive mean, but appearances can be deceptive. Also, the low resolution of the

Table 3.1. Properties of zero-mean Gaussian processes

Description	Symbol	Definition	Form for a Gaussian process	Remarks
Auto spectral density	$G(f)$	Mean square content of signal $X(t)$ per unit frequency f	Arbitrary	Frequency may be temporal or spatial. Gaussian behavior allows direct calculation of variance and expected zero-crossing rate
Variance	σ^2	Measure of signal dispersion	$\sigma^2 = \int_0^\infty G(f)\,df$	
Expected zero-crossing rate	$\bar{\nu}_0$	Average rate at which $X(t)$ crosses zero	$\bar{\nu}_0 = \frac{2}{\sigma}\sqrt{\int_0^\infty f^2 G(f)\,df}$	For a narrow-band process, $\bar{\nu}_0/2$ can be interpreted as an average sinusoidal frequency
Level exceedence curve	$E(x)$	Expected rate at which $X(t)$ crosses upward through level $x > 0$ or downward through level $x < 0$	$E(x) = \frac{\bar{\nu}_0}{2}\, e^{-x^2/2\sigma^2}$	$P(x) = -\,dE/dx$ for a narrow-band process $P(x) = \frac{\bar{\nu}_0 x}{2\sigma^2}\, e^{-x^2/2\sigma^2}$ for
Peak occurance curve	$P(x)$	Expected rate at which peaks $X(t) = x > 0$ or minima $X(t) = x < 0$ occur	Generally depends on ratio of total peaks to zero crossings, as well as σ, $\bar{\nu}_0$	a narrow-band Gaussian process. Narrow-band expressions are good approximations for broad-band $|x|/\sigma > 2$
Cumulative peak exceedence curve	$E_p(x)$	Expected rate at which peaks $X(t) > x$ or minima $X(t) < x$ occur	$E_p(x) = \int_x^\infty P(x)\,dx$ for peaks	

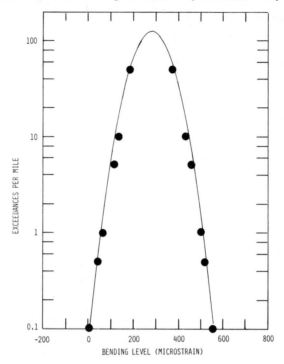

Figure 3.11.   Bending exceeding data and curve-fit for bus A-frame in a simulated revenue service test (strain gauge was located on leg of A-frame shown in Figure 3.5 (B)).

data makes it difficult to visually estimate the location of the mean to see if the exceedances are at least symmetric.

The curve shown in the figure is a Gaussian best fit that was estimated by means of a procedure which the author has found useful for dealing with low-resolution exceedance data. The steps in the estimation procedure are as follows.

First, pairs of positive and negative strains with equal exceedances are selected. (All of the data in Figure 3.11 is already paired; otherwise, a curve could be faired through the data points and pairs could be created by interpolation.) The average of a strain pair is considered to be a sample of the mean, and conventional statistical practice [12] is applied to estimate the mean value $\bar{X}$, the sample variance, and a confidence interval.

Second, the estimated mean is removed from the data and the adjusted sample is used to estimate the process variance. Let X_i be a typical data point and $Y_i = X_i - \bar{X}$ be the corresponding value with the mean removed. Now let $Y_i = k_i \sigma$, i.e., consider the adjusted data point to be located some number of standard deviations from the mean. This number, k_i, is unknown but can be estimated as follows. Select pairs of adjusted points Y_i, Y_j and let their exceedance be $E(Y_i) = n_i$ and $E(Y_j) = n_j$. Then, for a Gaussian distribution:

$$k_i = \sqrt{\frac{2 \log_n (n_j/n_i)}{1 - (k_j/k_i)^2}} = \sqrt{\frac{2 \log_n (n_j/n_i)}{1 - (Y_j/Y_i)^2}} \tag{3.12}$$

120

This procedure produces samples of the standard deviation $\sigma_i = Y_i/k_i$ and a sample size about half the number of original data points. The underlying standard deviation is then estimated as the sample average.

The final step in the procedure is to estimate the expected frequency. This is done by applying the Gaussian level exceedance formula (see Table 3.1) to one or more data points:

$$f_i = n_i \exp (k_i^2/2). \tag{3.13}$$

When working with equation (3.13), it is best to use the data points which lie closest to the mean.

Figure 3.12 shows two sets of exceedance data from the same bus undercarriage as in Figure 3.11. In this case, however, the data comes from runs made on a test track with a high density of artificial bumps and pot holes. The first data set (circles) is from the same strain gauge as in Figure 3.11. These exceedances may appear almost Gaussian to the eye, but they could not be adequately fit by the estimation procedure until a two-term Gaussian model was used. The curve-fit shown in the figure is:

$$E(\epsilon) = 69.1 \exp\left[-\frac{1}{2}\left(\frac{\epsilon - 333}{85.7}\right)^2\right] + 6.91 \exp\left[-\frac{1}{2}\left(\frac{\epsilon - 167}{52.3}\right)^2\right] \tag{3.14}$$

where ϵ and $E(\epsilon)$ are expressed in microstrain and exceedances per run, respectively.

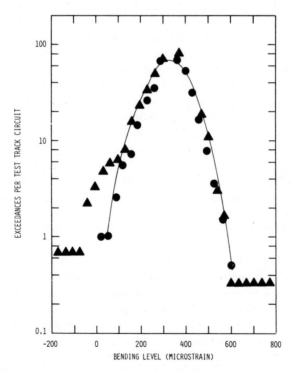

Figure 3.12. A-frame bending exceedances and one curve-fit from accelerated test on a 'torture' track.

121

The slight asymmetry of equation (3.14) suggests that the test-track strain-time history at this location can be treated as a weakly modulated process (see Figure 3.6). The likely source of the modulation is carbody motion, which the track excites to some small degree through the secondary suspension. Since the entire process has a narrow bandwidth, one can still expect the neighbouring peaks to be strongly correlated. It is then reasonable to reconstruct range-mean pairs from the data by taking:

$$\Delta \epsilon = \epsilon_{max} - \epsilon_{min} \qquad \epsilon_M = (\epsilon_{max} + \epsilon_{min})/2 \qquad (3.15)$$

for peak-strain pairs such that $E(\epsilon_{min}) = E(\epsilon_{max})$.

The second data set in Figure 3.12 (triangles) comes from a different strain gauge. In this case high strain levels are exceeded much more often than would be predicted by a simple Gaussian model, but the significant deviations appear only at very low exceedance rates. This behavior is typical for motion limited soft suspension systems: the strain is Gaussian most of the time but an occasional large motion engages a stop that increases the loads and strains imposed on the beam.

In this case the exceedance plot alone does not contain enough information to reconstruct the range-mean pairs because both positive and negative strain spikes are present. The spikes might occur in pairs if the suspension is lightly damped or the track inputs are sufficiently violent, or they might occur separately in a moderately damped suspension subjected to occasional violent inputs. The latter situation is likely for typical bus suspensions and roadways, and would suggest that positive spikes be combined with negative Gaussian levels and vice versa. Some supplementary counting of the time histories would be advisable, however, if the general dynamic properties of this particular suspension were not well known.

The foregoing example suggests that one might consider the motion-stop spikes as a separate random process which occasionally and briefly replaces the primary process in the time series. Therein lies a short cut that is also useful for treating aircraft as their operating characteristics change in various phases of flight. Consider, for example, what happens to the lower wing skin stresses in a transport airplane as it alternately cruises in calm air, encounters patches of moderate and severe turbulence, and finally reduces the wing bending moment by lowering inboard flaps on final approach to the runway. Counting a mission-length sample of such events would likely produce a non-Gaussian exceedance curve, but what has really happened is that the length of the sample has disguised a piece-wise Gaussian process. Once again, since the wing flexibility assures a narrow bandwidth, a reasonable summary of range-mean pairs can be read from the exceedance curve without having to sample each mission segment separately.

It is also essential to be aware of the possibility that changing operating characteristics might change the mean. This can happen in many ways, e.g., significant reduction of gross weight as a transport airplane consumes fuel on a long mission, asymmetric static suspension loads on railroad vehicles negotiating curves above or below balance speed, etc. In this case the reconstruction should include a separate Gaussian fit and separate identification of range-mean pairs for asymmetric humps in the occurrence curve (Figure 3.13).

Spectral analyses are also easy to convert to a load summary when dealing with narrow-band processes. The autospectral density function can be used to confirm the

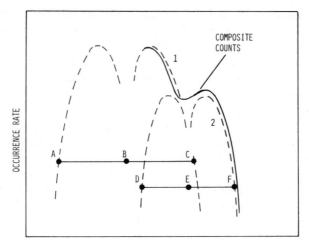

Figure 3.13. Method for identifying range-mean pairs in a composite occurrence curve covering different operating conditions.

OPERATING CONFIGURATION	RANGE	MEAN
1	AC	B
2	DF	E

bandwidth, look for modulations, etc. The formulas given in Table 3.1 for variance and frequency can be applied to non-Gaussian as well as Gaussian processes, with the understanding that the variance is a Gaussian best-fit value. For example, unquestioning acceptance of a Gaussian best fit for a suspension component subject to motion stops would overestimate the magnitudes of frequently occurring ranges and underestimate the magnitudes of infrequent ranges. One should also look at the probability density function, therefore, to determine if the error in the fit requires some adjustment in the load summary. Spectral analyses are similarly open to misinterpretation if the service loads change during a mission. However, these analyses are usually performed on short data samples, so the problem rarely arises in practice.

Strong modulation: the axle environment

Carbody static loads and their ground reactions establish four-point bending in the vertical plane and thus subject axles to rotating bending fatigue when automotive and railroad vehicles move. The service loading is much more complex than the analogous laboratory fatigue test, however, because guideway irregularities excite carbody and suspension structure resonances, power and brake applications induce axle bending in the horizontal plane, and pot holes in roads or interlockings in track cause impact loads that induce bending in both planes.

The magnitude of the dynamic axle bending tends to be about 10–20% of the static bending moment but can be of the same order as static bending for brief times during impacts. These are the characteristics of the strongly modulated process shown in Figure 3.6, the axle rotation acting as the modulator.

Although static bending is the major contributor to axle fatigue cycles, it is often necessary to establish a good separate summary of the dynamic effects. The need arises from the kinds of failure modes usually found in axles: fretting corrosion and fretting fatigue cracks which develop at interference fits such as bearings and drive gears. An accurate dynamic summary is required because small changes in the dynamic loads can sometimes make the difference between failure and survival when fretting at an interference fit is involved.

Axle dynamic environments pose a special problem in that they are difficult to deduce from single-channel measurements. First, the static load tends to be the dominant factor in the total environment. Second, the modulation distorts the measurements (e.g., consider the result if a single strain gauge happens to be passing through the horizontal bending plane just when the axle sees a vertical impact bending load).

The difficulty can be resolved by making two measurements at locations A, B separated by $90°$ around the axle circumference. The total bending magnitude is then obtained by processing the data to remove the modulation:

$$M = \sqrt{A^2 + B^2}. \tag{3.16}$$

In the demodulated signal, M, the static load appears as the mean value and the dynamic loads are the only contributors to the deviations. Sometimes three locations A, B, C at 120-degree intervals are used to provide redundancy in case of sensor failure. The demodulation procedure then takes the form,

$$M = \sqrt{A^2 + (B-C)^2/3} \tag{3.17}$$

and if (say) sensor A has failed, then:

$$M = \sqrt{(B+C)^2 + (B-C)^2/3}. \tag{3.18}$$

Figure 3.14 shows a typical example of single-channel and demodulated strain-time histories taken from a railroad axle as it was passing through an interlocking [14].

Fatigue loads can be specified directly from a demodulated load summary if the objective is to define a rotating bending laboratory test for service simulation. The service distribution of exceedances per material point can be approximately recreated by using slightly different frequencies for the loading and rotation.

If the objective is to predict life, however, reconstructing the load summary requires two more steps to account for the facts that a typical material point is not always in the principal bending plane when a load peak occurs, and that the point can be located so as to see positive loading when the axle dynamic load is either positive or negative.

As long as the wheel/axle set itself is not the load source,[*] one can assume that the dynamic peaks are uniformly distributed over the axle revolution. A simple expression

---

[*]For example, loads caused by tread defects on railroad wheels should be counted (conservatively) at one peak per revolution for the material point. One should first examine the autospectrum, therefore, to separate the one-per-revolution component of the RMS.

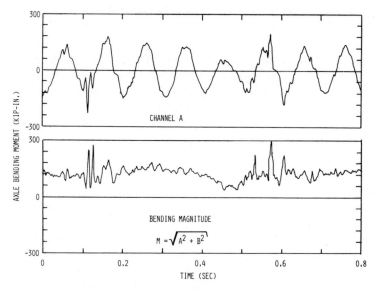

Figure 3.14. Bending moment versus time in a rotating axle (upper plot: single data channel showing rotational modulation; lower plot: the same time span showing demodulated signal obtained by processing two channels).

can then be derived for exceedances of the material-point dynamic load factor, m, defined by:

$$S_{\text{peak}} = (1 + m)S_{\text{static}}. \tag{3.19}$$

Suppose that the axle is subjected to a positive peak dynamic load factor $x \geqslant m$; then the positive level, m, will be exceeded at any material point lying within the critical angle:

$$\theta = \cos^{-1}(m/x) \tag{3.20}$$

from the positive principal bending plane. Negative dynamic peaks will similarly reduce the level exceeded. These two effects are reversed for material points within $\pm \theta$ from the negative principal bending plane, where the static load contributes negative bending.

The first extra step is to obtain the level exceedance curve for the material point by summing axle peak occurrences weighted by the fraction of time the point spends in the critical zone:

$$E(m) = \frac{1}{\pi} \int_m^\infty \cos^{-1}(m/x_{\text{max}})P_{\text{axle}}(x_{\text{max}})dx_{\text{max}} \quad (m > 0)$$

$$\tag{3.21}$$

$$E(m) = \frac{1}{\pi} \int_m^\infty \cos^{-1}(m/x_{\text{min}})P_{\text{axle}}(x_{\text{min}})dx_{\text{min}} \quad (m < 0).$$

125

The peak occurrences $P(m)$ are easily obtained by taking differences:

$$P(m)\Delta m = E(m) - E(m + \Delta m). \tag{3.22}$$

These level exceedances and occurrences are distributed about the positive static stress level. Also, note that equations (3.21) must be reflected to describe the distribution about the negative static stress level. If the axle dynamic loading is a Gaussian process, $P(x_{\min}) = P(x_{\max})$ and equations (3.21) reduce to:

$$E(k) = \frac{\bar{f}}{\pi} \int_k^\infty \xi \cos^{-1}\left(\frac{|k|}{\xi}\right) e^{-\xi^2/2} \, d\xi \tag{3.23}$$

for both positive and negative k, where $m = k\sigma$, $x_{\min} = x_{\max} = \xi\sigma$ and σ^2 is the process variance. Figure 3.15 compares the nondimensional level exceedance curves for the whole axle and the typical material point.

Finally, the static and dynamic loads must be recombined. The static load is a fatigue cycle for the material point; if no other effects are active, it has range $\Delta S = 2S_{\text{static}}$, a zero mean, and an occurrence rate $f_0 = V/\pi D$, where V is the vehicle speed and D is the wheel or tire diameter.[*] The ratio of the dynamic process frequency to f_0 affects the recombination. The following procedures are suggested:

(1) *Carbody dynamic environment* $(\bar{f} \ll f_0)$. The dynamic effects act as a modulator. For the range-mean pair,

$$\Delta S = 2(1 + k\sigma)S_{\text{static}}, \qquad S_M = 0 \tag{3.24}$$

estimate the occurrence rate as:

$$f(k) = \frac{f_0 E_{\text{axle}}(k)}{\int_{-\infty}^\infty E_{\text{axle}}(k)dk} \tag{3.25}$$

using the whole-axle level exceedance curve, and group cycles in bands $f(k)\Delta k$. The denominator in equation (3.25) normalizes $f(k)$ so that the sum of occurrences of all ranges equals f_0. For a Gaussian process equation (3.25) becomes:

$$f(k) = (f_0/\sqrt{2\pi})e^{-k^2/2}. \tag{3.26}$$

(2) *Comparable frequencies* $(\bar{f} \approx f_0)$. This case represents both low-frequency suspension structure resonances and carbody dynamics at low speeds. For the range-mean pairs defined by equation (3.24), estimate the occurrence rate approximately by using the weighted level-exceedance curve of equation (3.21) in place of $E_{\text{axle}}(k)$ in equation (3.25). For a Gaussian process this reduces to:

$$f(k) = 0.402 \int_{|k|}^\infty \xi \cos^{-1}\left(\frac{|k|}{\xi}\right) e^{-\xi^2/2} d\xi \tag{3.27}$$

or $0.7925f(k)$ can be read directly from the demodulated exceedance curve in Figure 3.15.

---

[*]It is sometimes convenient to work with spatial occurrence rates, e.g., $f_0 = 63360/\pi D$ per mile with D expressed in inches or $f_0 = 1000/\pi D$ per km with D expressed in meters.

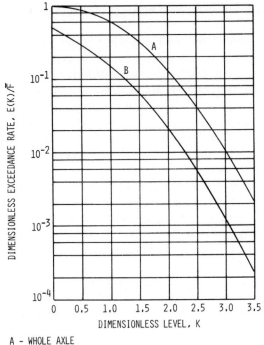

A - WHOLE AXLE

B - TYPICAL MATERIAL POINT

Figure 3.15. Whole-axle and material-point level exceedance curves for a rotating axle subject to a narrow-band Gaussian process.

(3) *High-frequency dynamics* $(\bar{f} \gg f_0)$. In this case transient responses to impacts and suspension structure resonances at higher frequencies occasionally increase or decrease either a positive or negative peak (but not a pair) in the static load cycle. Therefore, define the range-mean pair as:

$$\Delta S = (2 + m)S_{\text{static}}, \qquad S_M = mS_{\text{static}}/2 \qquad (3.28)$$

and use $P(m)\Delta m$ from equations (3.21) and (3.22) to estimate the occurrence rate. Note that $P(m)\Delta m$ implicitly includes the dynamic frequency, as it should in this case. Also note that range-mean pairs $(\Delta S, S_M) = (2S_{\text{static}}, 0)$ must be included in the load summary at the adjusted rate of occurrence $f_0 - \Sigma P(m)\Delta m$.

Broadband loading: maneuver-dominated structure

Maneuver loads are generally broadband processes because there is no isolation filter between the maneuver commands and the loads they cause. This poses no problem for civil transportation vehicles because the significant maneuvers tend to be infrequent and quasi-steady, and the remainder tend to be submerged by the other narrow-band processes that are active. Fighter aircraft are in a special category, however, because maneuver loads make the dominant contribution to stresses in many fighter airframe components.

Consider, for example, a lower skin stress in a fighter wing. The effects of maneuver loads on this stress can be represented to a first approximation by following the history of acceleration normal to the wing planform at the airplane center of mass. The acceleration is commonly reported in terms of load factor, n, i.e., the number of 'g's the pilot is pulling, with $n = 1$ defined as steady level flight. Figure 3.16 presents schematic n-time histories for several common fighter maneuvers. Approximate 'g' levels have been noted, and the figure includes a similar plot for gust response, a ground-air-ground cycle, and a composite exceedance curve for a typical fighter mission.

The lessons to be drawn from Figure 3.16 are that maneuver loads are by far the most important source of fighter wing fatigue stress, that the positive and negative maneuver peaks combine asymmetrically in both magnitude and occurrence rate, and that no obvious way exists to reconstruct a fatigue-load summary from a fighter exceedance curve alone. The U.S. Air Force bases reconstructions on detailed mission/ segment analyses [15].

The broadband Gaussian process has been used to generate summaries by first transforming the positive and negative exceedance curves to the same zero-mean unit probability distribution, then pairing positive and negative peaks by means of a Monte Carlo simulation, and finally transforming back to the original variable [16]. This procedure must be applied directly to wing skin stress rather than the load factor. Several exceedance curves are required to reflect the change in relation between stress and load factor as the airplane is reconfigured, e.g., by deployment of flaps, leading-edge slats, or dive brakes, or by accelerating from subsonic to supersonic flight. The Monte Carlo simulation must, in essence, follow the airplane through its mission and must include weighting factors to create realistic peak pairs based on how pilots handle the airplane.

The example of the fighter airplane wing highlights the most important limitation of shortcut methods for loads analysis. When dealing with structures that are truly broadband-loaded, there is no substitute for detailed knowledge. The only way to get that knowledge is to build a picture of each type of maneuver, including correlations with range-pair or rainflow strain counts at representative locations in the structure. Only after such information has been developed can one usefully apply methods like the Monte Carlo simulation described above.

Limitations of accelerometer data

No discussion of service loads analysis should be ended without some remarks on the interpretation of accelerometer data. The use of counting accelerometers to track load-factor exceedances in military airplanes was mentioned earlier. The airplane load factor is especially useful when dealing with fighter wings because the stress response to maneuvers is essentially static. Load factor is less useful for a transport airplane wing because wing flexibility tends to isolate the airplane center of mass from gust loads.

Accelerometers are also employed in ground-vehicle tests, where they are useful

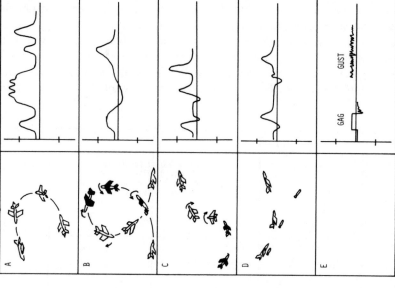

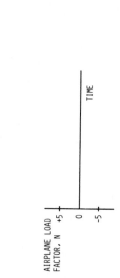

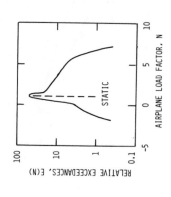

LEGEND:

(A) COMBAT TURNS (SECOND TURN IN N-T PLOT ILLUSTRATES STALL-BUFFET CHARACTERISTIC AT MAXIMUM LIFT COEFFICIENT)

(B) CLIMB AND ROLL INTO OUTSIDE LOOP FOLLOWED BY PULLOUT

(C) COMBAT TURNS WITH DISENGAGE/ENGAGE SNAP-ROLLS

(D) DIVING APPROACH TO GROUND TARGET, STORES RELEASE, AND PULLOUT

(E) GROUND-AIR-GROUND CYCLE AND EFFECTS OF GUSTS DURING CRUISE

Figure 3.16. Typical load-factor environment for a fighter airplane.

for assessing ride-quality factors and external loads on the suspension. Both of these quantities are associated with rigid motions of the carbody. However, there is also a tendency in this field to commit crimes against Newton's second law by misinterpreting high-frequency acceleration data in connection with internal suspension loads and loads imposed on unsprung masses. Two brief examples will be given to illustrate the proper interpretation of acceleration data in these environments.

First, it is a good general rule that high peak accelerations occur only at high frequencies. Figure 3.17 summarizes a numerical experiment that illustrates this characteristic by means of several analyses in which raw acceleration data was filtered at different cutoff frequencies before exceedances were counted. The raw data in

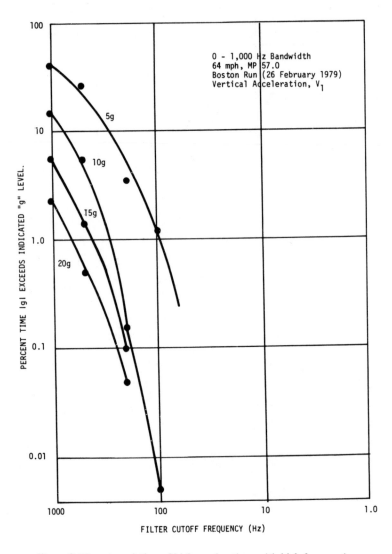

Figure 3.17.   Association of high accelerations with high frequencies.

this case came from the bearing on a railroad axle, and had significant mean-square content out to 700 Hz [13]. The results plotted in Figure 3.17 show that the high-'g' peaks tend to disappear when the high frequencies are filtered out of the signal.

Second, the significant stresses in a dynamically excited system are generally associated with rigid-body and low-frequency vibration modes. To illustrate this point, gearbox accelerations have been compared with bending strain measurements made on a powered railroad axle [14]. Figure 3.18 contrasts the two autospectra: significant mean-square content out to 400 Hz in the acceleration but only 100 Hz in strain. Figure 3.19 plots the ordinary coherence function [12] between the two measurements, showing that they are not related above 100 Hz.

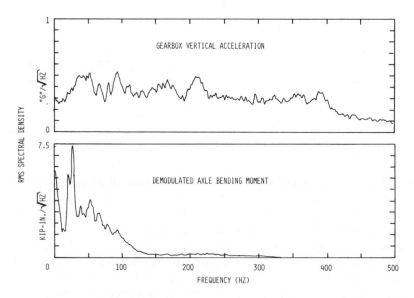

Figure 3.18. Difference between acceleration and bending responses of a railroad axle (the ordinates of these plots must be squared to obtain the autospectral densities).

These two cases should serve as a warning that accelerometer data is not to be trusted in suspension and unsprung environments without elaborate measurement coverage and accurate knowledge of flexible mode shapes for calculation of generalized masses. When working with such environments the best course is to rely on strain-gauge data.

3.5. Predicting safe life

The railroad wheel failure mentioned in the second section was an isolated case. Management action was all that was required to avert a repetition once the cause had been identified as poor enforcement of inspection procedures.

The proper course of action is harder to discern and the risks are greater, however, when a fleet is genuinely in a state of incipient structural failure. Such states can result

131

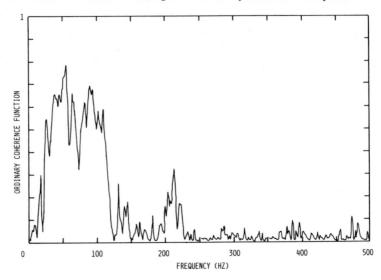

Figure 3.19.   Correlation between axle acceleration and bending (this plot is for the same data as in Figure 3.18. A coherence value of one means perfect correlation; a zero value means no correlation).

from underdesign, overusage, or aging — causes which have all affected various military [17] and civil [18] airplane fleets as well as ground transportation vehicles [3, 13, 14] and structures [19, 20]. Dealing with fleet failures requires damage analysis based on load summaries like those discussed in the preceding sections.

The first step in applying damage analysis to an incipient fleet failure problem is to adjust the ground rules and procedures until the life predicted for the existing structure agrees reasonably well with early-failure data from the fleet. The safe life for modified structure or modified usage can then be predicted and compared with the service life requirement. This approach generally does not demand a precise model for damage accumulation as long as the ground rules limit the relative predictions to similar structure geometry, material behavior, and service environment.

Linear damage summation satisfies the foregoing criteria and (as will be seen shortly) provides the means for rapid analyses that highlight the sensitivity of safe life to modifications. Linear damage summation is thus an ideal method for most failure investigations, although some important exceptions will be noted in the following discussion.

Linear damage summation

Linear damage summation was first proposed as an empirical method for predicting service fatigue life from *S–N* curves obtained from laboratory standard constant-amplitude tests [21, 22]. Early applications involved low service stress amplitudes, a regime where crack nucleation dominates the lifetime. The procedural rules for linear damage summation are as follows:

(1) For an alternating stress exceeding the endurance strength damage is linearly proportional to the number of stress cycles accumulated.
(2) The fully reversed bending ($S_M = 0$; $R = -1$) $S-N$ curve determines the relative rates of material damage caused by alternating stresses with different amplitudes.
(3) The damage rate is adjusted by means of a modified Goodman diagram or equivalent plot for cycles with mean stress not equal to zero.
(4) The damage rate does not depend on the sequence in which different stress cycles occur.

In the most widely used form of the procedure, rules (2) and (3) above are mediated by the number, N, of constant-amplitude/mean stress cycles to failure, i.e., the damage D is expressed as:

$$D = \sum_i (n_i/N_i) \qquad (3.29)$$

where n_i cycles of the ith stress condition (S_{Ai}, S_{Mi}) are applied and N_i such cycles under fixed stress conditions would have caused failure, based on laboratory $S-N$ data. Predicted life is then expressed as $L = 1/D$ and is interpreted as the number of repetitions or blocks of the 'spectrum' (n_i, S_{Ai}, S_{Mi}; for all conditions $i = 1, 2, \ldots$) that should cause failure.

For a random-load spectrum the average damage rate is defined by:

$$D = \int\int \frac{P(S_A, S_M)}{N(S_A, S_M)} \, dS_A \, dS_M \qquad (3.30)$$

where $P(S_A, S_M)$ is the joint amplitude-mean occurence density function and $1/D$ is interpreted as the expected lifetime. For example,

$$D = \frac{\bar{f}}{\sigma^2} \int_0^\infty \frac{S_A e^{-S_A^2/2\sigma^2}}{N(S_A)} \, dS_A \qquad (3.31)$$

if the loading is a zero-mean narrow-band Gaussian process with expected frequency $\bar{f}$ and variance σ^2. Expressions like equation (3.31) are easily integrated analytically when $S-N$ data is fit with power laws of the form $(S_A)^r N = $ constant in the mid to high cycle fatigue regimes.

The linear damage summation method has been extended to combined-stress situations by assuming that the uniaxial-stress $S-N$ diagram applies directly to octahedral stress [23]. Also, a supplementary model for dealing with cumulative reduction of endurance strength caused by stresses above the endurance limit has been correlated with steel-alloy fatigue test data [24]. Linear damage summation has been used for over thirty years to design fatigue-resistant structures, and a large body of design practices for many applications now exists [25].

Despite this extensive practical application, reliable prediction of crack-nucleation life remains elusive in many cases. Actual life is highly scattered in constant-amplitude tests at stresses near the endurance strength. Crack propagation begins to represent a significant fraction of total life in the mid to low cycle fatigue regime. Fretting fatigue at interference-fit attachments cannot be analyzed without laboratory tests conducted specially to generate *S–N* data for the specific configuration of interference and fastener flexibility because the physical behavior under these conditions differs from fretting action in standard laboratory tests. The stress raisers in a structure can cause small-scale plastic yielding under elastic nominal stresses, making the crack-nucleation process dependent on the fatigue cycle sequence [26]. In combined-stress situations the distribution of physical damage (and hence the damage rate) can change in unknown ways as the phase relations between stress components change.

Some progress has been made recently in the treatment of small-scale plasticity effects; also, the replacement of *S–N* data with strain-*N* data from strain-controlled tests in which *N* is related to the appearance of a small crack with specific size has reduced the scatter in the test data [5]. Crack-nucleation life prediction retains a fundamental limitation, however, in that the damage rate *D* has no physical meaning to relate to micromechanical models of the fatigue damage process.

Fretting corrosion and abrasive wear

Failures involving fretting corrosion or abrasive wear are at least potentially susceptible to rational analysis based on linear damage summation. The rates of these processes can be quantified in terms of material volume loss from the contacting surfaces, and the end of useful life can be specified in related terms of condemning limits on part tolerances.

The damage rates for both processes are linearly proportional to the contact pressure and also depend on the amplitude of relative motion between the contacting parts. Abrasive wear rates tend to be linearly proportional to motion amplitude, but fretting corrosion rates generally have a threshold characteristic [27], which can be adequately modelled with a bilinear dependence. Neither process is affected by the sequence of loads or motions.

The general characteristics of abrasive wear and fretting corrosion fit the linear damage summation hypothesis. Furthermore, the relation between cumulative damage and cumulative motion is linear when the amplitude dependence is linear, i.e. a spectrum damage rate can be calculated by summing the positive and negative peak occurrences of mean-removed motion. For a zero-mean random-motion process, $x(t)$, and a constant contact pressure, p, the volume loss rate is thus given by:

$$D = Cp\left[\int x_{max} P(x_{max})\mathrm{d}x_{max} + \int x_{min} P(x_{min})\mathrm{d}x_{min}\right] \tag{3.32}$$

where the constant of proportionality, C, is a property of the material combination and the contact geometry. If $x(t)$ is a narrow-band Gaussian process with expected frequency $\bar{f}$ and variance σ^2, then equation (3.32) reduces to:

$$D = \sqrt{2\pi}\, C\bar{f}p\sigma. \tag{3.33}$$

Predicted life is simply $L = D_{cr}/D$, where D_{cr} is the condemning limit expressed in terms of volume loss.

Fretting corrosion damage analysis can also be applied to interference-fit attachments, in which both contact pressure and motion amplitude depend on accumulated damage. Experiments are generally required to determine the relationships $p(D)$ and $x(D)$ under constant-amplitude external loading. One can then replace the constant-pressure factor by:

$$p(x) = p[x^{-1}(D)] \tag{3.34}$$

under the integral signs in equation (3.32) to estimate the random-load damage rate.

The ability to apply damage analysis to wear and fretting situations is restricted by limitations on measurements rather than the linear damage summation hypothesis. For example, the relative motions associated with fretting corrosion are in the range of 10^{-3} to 10^{-2} in (0.02 to 0.2 mm) and are difficult to measure accurately in an externally loaded interference-fit experiment. One might still be able to perform a comparative analysis using a proxy variable (e.g., bending moment on an axle with interference-fit bearings), except that the transition amplitude is unknown. The transition threshold cannot be determined from small-scale standard laboratory tests because the service contact geometry is different and has a different threshold. Also, it is generally impractical to directly determine the fretting rate from a laboratory test of service equipment because the associated weight and dimensional losses are smail.

Crack propagation

Estimates for the safe lives of propagating cracks are based on empirical growth rate data [1, 2] which generally has low scatter. For example, Figures 3.20 and 3.21 illustrate the repeatability of crack growth rates in plain carbon rail steel. Two independent investigators [28, 29] made these measurements on samples taken from

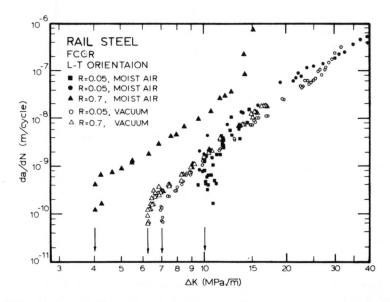

Figure 3.20. Rail-steel fatigue crack growth rates in air and vacuum (reproduced from Ref. [28]).

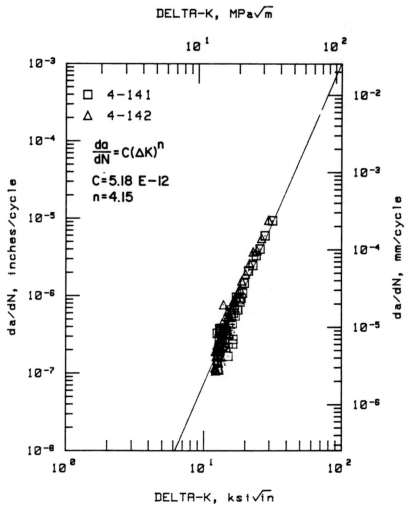

Figure 3.21.   Rail-steel fatigue crack growth rate in air at $R = 0$ (reproduced from Ref. [29]; compare with moist air data at $R = 0.05$ in Figure 3.20).

two different railroad rails. The repeatability evident in the figures is typical for crack growth rate data.

Estimating crack-propagation life requires a procedure for combining the effects of different cycles based on growth rates that were measured at essentially constant conditions.* The following hypothesis outlines the procedure which is followed in applied fracture mechanics:

(1) The crack-length increment per cycle, da/dn, can be represented in terms of a

---

*Most laboratory tests for crack growth rates are run at constant load, so that the stress intensity range ΔK increases as the crack length increases. The variation of ΔK is negligible, however, over the individual crack-length increments from which the growth rate data points are calculated.

continuous functional relationship with the stress cycle parameters, i.e., $da/dn = F(\Delta K, R)$.

(2) There exists a threshold for the stress intensity range, ΔK, below which cracks do not grow. This threshold is represented by K_{th} at $R = 0$, but its value generally depends on the stress ratio, R.

(3) A large ΔK can reduce the crack growth rate during subsequent cycles at smaller ΔK (the 'retardation' or 'load interaction' effect).

(4) For structure/crack geometries and remote stress distributions other than the test conditions, the same rate equation $da/dn = F(\Delta K, R)$ applies with ΔK calculated from an appropriate stress intensity (K) formula derived from linear elastic fracture mechanics.

(5) Life is determined by summing the crack-length increment per cycle between crack-length limits specified on the bases of detectability and loss of structural integrity. For long lives the crack-length increment sum can be approximated by integrating the rate equation as if it were a continuous first-order differential equation.

Over fifty rate equations have been proposed to fit curves to the empirical data. The most widely used is the so-called Forman equation [30]:

$$da/dn = C(\Delta K)^r/[(1 - R)K_C - \Delta K] \tag{3.35}$$

which models the mid and high ΔK regimes (see Figure 3.20 or 3.21). The parameters C, r are chosen to fit the mid ΔK data, and K_C is a fracture toughness parameter chosen to fit the high ΔK data where the condition of fracture is approached. In some applications most of the ranges occur in the mid and high ΔK regimes and the threshold is treated simply as a sharp cutoff, e.g., $da/dn = 0$ for $\Delta K/(1 - R) = K_{max} \leqslant K_{th}$. In other applications most of the ranges occur at low to mid ΔK; it is then better to ignore the high ΔK regime and model the other two regions directly, e.g.:

$$da/dn = C[\Delta K - (1 - R)K_{th}]^r \tag{3.36}$$

which allows for a gradual growth rate cutoff at low ΔK.

In either kind of application, it often happens that the stress intensity range occurrences occupy only a small region on the $da/dn - \Delta K$ diagram, and a simple power law:

$$da/dn = C(\Delta K)^r/(1 - R) \tag{3.37}$$

can then be fit to the region of interest. Figure 3.22 is an illustrative example taken from a crack growth analysis of a beam in a bus undercarriage [3].

Load interaction effects have been observed in the mid to high ΔK regimes in tests involving typical aircraft materials and fighter stress spectra [31, 32]. The major effect appears to be retardation of the crack growth rate and is accounted for by various supplementary algorithms which modify the basic rate equation during the crack-length increment summation. The modifications are usually implicit, i.e., the form of the basic equation is retained while the current ΔK is modified in accordance with running calculations based on ΔK or K_{min}, K_{max} values from preceding cycles [33, 34]. Load interaction effects should similarly be expected when the stress intensity ranges are concentrated near the crack-propagation threshold but the low

137

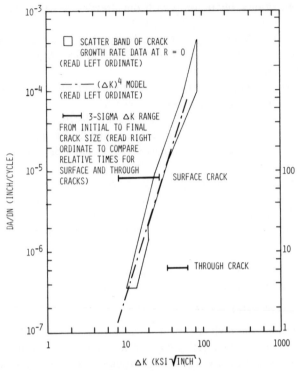

Figure 3.22.   Simple power-law model for crack growth rate equation (used to analyze a detail of the bus A-frame shown in Figure 3.5 (B)).

growth rates in this case (of the order of one atomic lattice spacing per cycle) involve non-continuum mechanisms, and no general physically based rate models have yet been developed to treat this regime [35].

Stress intensity factor fomulae based on singularity solutions of boundary-value problems are available today for a wide variety of structure/crack geometries and remote stress distributions [36]. These formulae can be applied to crack growth calculations in the general form:

$$\Delta K = G(a)\sqrt{a}\ \Delta S \tag{3.38}$$

where $G(a)$ is a dimensionless function of the crack length, a, and other parameters. The function $G(a)$ accounts for local geometric influences (e.g., the radius of a fastener hole from which the crack is growing) and nonuniformities in the remote stress distri-bution (e.g., a linearly varying bending stress with ΔS representing the stress range at the extreme fibers of the section). The available K-solutions include three-dimensional cases for which singularities can be defined (e.g., elliptical cracks embedded in solids [37] and correction factors for surface cracks [2]. Applying these formulae to a practical case, however, is not always a simple matter. The cracks actually observed in structures often have complex shapes (particularly in the three-dimensional cases) which violate the assumptions, implicit in the singularity solutions, that the crack remains a plane surface and the crack-front shape does not change during propagation.

138

Rapid analysis of such cases requires selection of the most nearly applicable crack geometry for which K-solutions are available, but such judgments should be confirmed by test.

The objective of predicting crack-propagation life is usually to determine the maximum safe interval between periodic service inspections for crack detection and repair. Hence, the initial crack length should represent a crack that can be reliably detected by the inspection equipment and procedures which will be used in the particular application. When service inspection is impractical the initial length must be based on manufacturer inspection and the safe interval should have an ample margin beyond the end of service life. The importance of confident knowledge about the capability of the actual inspection technique cannot be overemphasized; life predictions are most sensitive to the initial crack length and are meaningless if based on an unfounded estimate.

Loss of structural integrity through fracture or net-section rupture determines the end of the safe crack growth interval. The life of a high-strength material is usually limited by a fracture toughness parameter, K_C, when the crack becomes long enough so that $K_{max} = K_C$ on the next load cycle. For low-strength materials with high ductility the zone of plastic deformation ahead of the crack front can approach a size of the order of the remaining uncracked ligament before fracture occurs. Strictly speaking, one should apply ductile fracture mechanics to such situations. However, for life-prediction purposes it is reasonable to assume that the limiting crack length corresponds to net-section yield, i.e. when $S_{max} A = S_Y A_{net}$ where S_Y is the material yield strength and A, A_{net} are the total (original) and net uncracked cross section areas, respectively. For random service loading the final crack length can be determined via the implied static residual strength relation (either fracture or rupture) in terms of a first-passage analysis. Since one is interested mainly in extreme stress passages for the present purpose, it is reasonable to assume that the events are isolated and Poisson-distributed in time, leading to:

$$T = 1/E(S_{max}) \tag{3.39}$$

as the expected time to first passage [38], where $E(S_{max})$ is the level exceedance curve. Thus, if T is set equal to an inspection interval and S_{max} is obtained from equation (3.39), then S_{max} is expected to occur once per inspection interval.[*]

Numerically summing the rate equation block by block is currently the only reliable method for predicting the life of a propagating crack subject to significant retardation effects. A short cut can be taken in all other cases, however, by working with the integrated form of the rate equation. For example, the integrated forms corresponding respectively to equations (3.35) to (3.37) can be expressed as:

$$\frac{1}{N} = \frac{C(2S_M + \Delta S)(\Delta S)^{r-1}}{\int_{a_0}^{a_f} \dfrac{2K_C - (2S_M + \Delta S)G(a)\sqrt{a}}{[G(a)\sqrt{a}]^r}\, da} \tag{3.40}$$

---

[*]It may be advisable to use a time longer than the intended inspection interval to define S_{max} because of the uncertainty associated with the expectation, viz [38]: $\sigma_T = T$.

$$\frac{1}{N} = \frac{C(\Delta S)^r/(2S_M + \Delta S)^r}{\int_{a_0}^{a_f} \frac{da}{[(2S_M + \Delta S)G(a)\sqrt{a} - 2K_{th}]^r}} \tag{3.41}$$

$$\frac{1}{N} = \frac{C(2S_M + \Delta S)(\Delta S)^{r-1}}{2\int_{a_0}^{a_f} \frac{da}{[G(a)\sqrt{a}]^r}} \tag{3.42}$$

where $1 - R = 2\Delta S/(2S_M + \Delta S)$ and equation (3.38) have been introduced. Equations (3.40) to (3.42) are essentially *S–N* diagrams that specify the number of constant-range/mean cycles, N, required to grow a crack from the initial length a_0 to the final length a_f. It can easily be shown that equation (3.37) exactly satisfies the linear damage summation hypothesis, i.e., the life prediction $L = 1/\Sigma(n_i/N_i)$ with N_i calculated from equation (3.42) produces the same results as block by block numerical summation. Equations (3.35) and (3.36) depend on the range/mean sequence. However, a recent numerical experiment has shown that linear damage summation can be applied to the integrated forms of these models (equations (3.40) and (3.41)) and that this short cut will produce life estimates reasonably close to the exact block-by-block results for practical cases where the lifetime contains many spectrum blocks [39].

Predicting unretarded crack growth life under random loading now becomes a simple matter of summing over the appropriate stress occurrence density function. For example, a narrow-band Gaussian spectrum with variance σ^2 on the amplitude and expected frequency $\bar{f}$ leads to the following life expressions for the Forman model (equation (3.40)) and the plain power-law model (equation (3.42)), respectively:

$$\frac{1}{L} = (2\sigma)^{r-1}C\bar{f} \int_0^\infty \frac{\left(\frac{S_M}{\sigma} + x\right)x^r e^{-x^2/2} dx}{\frac{K_C}{\sigma}\gamma(a_0, a_f, r) - \left(\frac{S_M}{\sigma} + x\right)\gamma(a_0, a_f, r-1)} \tag{3.43}$$

$$\frac{1}{L} = \frac{(2\sqrt{2}\sigma)^r C\bar{f}}{2\gamma(a_0, a_f, r)}\left\{\Gamma\left(\frac{r+2}{2}\right) + \frac{S_M}{\sigma\sqrt{2}}\Gamma\left(\frac{r+1}{2}\right)\right\} \tag{3.44}$$

where $\Gamma(r) = \int_0^\infty x^{r-1}e^{-x}dx$ is the Gamma function and:

$$\gamma(a_0, a_f, r) = \int_{a_0}^{a_f} \frac{da}{[G(a)\sqrt{a}]^r} \tag{3.45}$$

is the crack-geometry integral. The gradual threshold model (equation (3.41)) must be left as a double summation or integration, however, except for special cases where the growth rate exponent, r, is an integer.

Life expressions such as equations (3.43) and (3.44) are convenient for analyzing piece-wise sequential random loading described by several different sets of occurrence statistics. These formulae are also convenient for analyzing multi-stage crack growth, in which the propagation is followed through one or more components of a built-up section and one or more transitions from surface crack to through crack. In general, a crack-geometry integral $\gamma(a_{0i}, a_{fi}, r)$ must be calculated for each stage $i = 1, 2, \ldots, I$. The life prediction for a multi-stage/multi-spectrum analysis is given by:

$$L = \sum_{i=1}^{I} \left(1 \Big/ \sum_{j=1}^{J} L_{ij} \right)^{-1} \tag{3.46}$$

where L_{ij} is the life in the ith stage under the jth spectrum.

Sharp-cutoff threshold models can be treated as a special case of multi-stage crack growth by choosing several intermediate crack lengths to subdivide the total life into $(a_0, a_1, a_2, \ldots, a_f)$. In the ith stage (a_{i-1}, a_i) the stress spectrum is simply truncated below:

$$(2S_M + \Delta S)_i = 2K_{th}/G(a_{i-1})\sqrt{a_{i-1}}. \tag{3.47}$$

For example, the Forman and plain power-law narrow-band Gaussian spectrum lives with a sharp threshold cutoff are given respectively by:

$$1/L_i = (2\sigma)^{r-1} C\bar{f} \int_{x_i}^{\infty} \frac{\left(\dfrac{S_M}{\sigma} + x\right) x^r e^{-x^2/2} dx}{\dfrac{K_c}{\sigma} \gamma(a_{i-1}, a_i, r) - \left(\dfrac{S_M}{\sigma} + x\right) \gamma(a_{i-1}, a_i, r-1)} \tag{3.48}$$

$$1/L_i = \frac{(2\sigma)^r C\bar{f}}{2\gamma(a_{i-1}, a_i, r)} \int_{x_i}^{\infty} \left(\frac{S_M}{\sigma} + x\right) x^r e^{-x^2/2} dx \tag{3.49}$$

where

$$x_i = \frac{K_{th}}{\sigma G(a_{i-1})\sqrt{a_{i-1}}} - \frac{S_M}{\sigma} \tag{3.50}$$

Confirmation testing

Damage analysis is most useful for assessing the life of a modification when the life predicted from conservative assumptions has an ample safety margin over service life. However, it is sometimes impossible to make unambiguous predictions (e.g., for fretting at interference fits), and in other cases the predicted safety margin may be less than the uncertainty associated with the analysis assumptions. Such situations demand confirmation testing if further modification is impractical, but in general the circumstances of a failure investigation also impose constraints on testing time and resources. The following remarks are intended only to sketch the major options and to contrast confirmation testing with the more elaborate tests appropriate in research and development programs.

(1) *Observation of high-time service articles.* Even if other tests are conducted, observing the modified fleet is always advisable because it provides cheap additional insurance. If no other tests are conducted, watching the high-time articles preserves the structural integrity of the fleet by allowing some reaction time but risks the large economic loss of a second retrofit program if the first modification proves to be inadequate.

In the case of the failing railroad disc brake attachments mentioned earlier, [13] the operator decided to retrofit the fleet with alternate hardware supplied by different manufacturers at a time when only limited field-trial observations were available on a small number of samples. These samples had been clearly marked, however, and could subsequently be followed as high-time items. Also, only limited visual inspection was needed to determine if a disc brake was failing, i.e., the high-time items could easily be watched within the normal fleet maintenance schedule. In this case a good gamble of cheap fleet observation against high laboratory test costs paid off.

The principal issues one must face when attempting to rely on fleet observation alone are the ability of the maintenance system to track individual components in the fleet and the question of when to inspect. Good scheduling is vital when inspections entail significant extra costs or downtime. Damage analysis should be brought to bear on scheduling, even if the estimates are uncertain.

(2) *Accelerated life testing.* This option is frequently employed in failure investigations when the test cost is modest and the test can be completed before major modification funding must be committed. These criteria require that the accelerated life test be much less elaborate than R&D fatigue testing.

The railroad axle bending loads analysis mentioned earlier [14] is part of an effort to resolve a bearing fretting problem. The next step will be to subject an original and a modified axle to a rotating-bending fatigue test based on a summary of the measured loads. The test option is advantageous in this case because only a few thousand dollars and one to two weeks of test time have to be weighed against a retrofit cost of the order of one million dollars.

The principal issue one faces when considering an accelerated life test is spectrum truncation. The cost/benefit balance usually demands much more truncation in an accelerated life test than in an R&D fatigue test, often to the point of a single overload amplitude. Damage analysis should be used if possible to guide the test specification via estimates of the distribution of damage rate (n_i/N_i) across the spectrum. The degree of truncation and the uncertainty of the analysis demand that an accelerated life test always begin with the original article to verify test realism by reproducing the service failure mode.

(3) *Proof testing.* High fleet downtime and high direct costs make proof testing the option of last resort. Proof testing should not be considered unless the need to keep a fleet operating far outweighs the cost of testing every vehicle and the risk of destroying a few in the process.

Two proof test programs were carried out in the 1970's on military airplane fleets [17]. In each case the fleet had to be kept flying to maintain a strategic mission capability, and proof testing was the only means to do so while waiting for the production of retrofit kits.

The proof test is essentially a static test of the structure's ability to carry its maximum service load without failing. The principal issue involved in the specification of a proof test is the possibility that the test load may fatally damage the structure without causing immediate failure or giving other obvious evidence of the damage. This is a problem peculiar to crack-propagation damage. If the critical components are made of steel alloys which have a significant fracture-energy transition below the lowest expected service temperature, the proof test load can be kept below damaging levels by cold-soaking the structure before applying the load. If the critical components do not have such a transition, the proof test must be accompanied by subsequent operational and payload restrictions to reduce the service loads.

3.6. Maintaining perspective

Failure mechanics resembles the broader system-level practice of risk assessment. Others have said that risk assessments can become meaningless if analysts overemphasise those aspects of risk which fit neatly into the calculus of probability. The same can be said about failure mechanics if analysts pay too much attention to damage accumulation models.

The short introductory essay on the fractured railroad wheel incident was intended to serve as a warning against the attitude of trying to make the problem fit the model. The sequel having been devoted to discussion of models, it is now time to repeat and extend the warning by means of some brief sketches of other 1970–80 cases.

The importance of recognizing the difference between structural failure and post-accident damage is illustrated by a case of a light airplane crash in which a failure-investigation consultant wrongly attributed the cause to understrength material in the airplane's aluminum propeller. The error arose from examination of micrographs of a sample taken from material near a fracture which probably happened when the propeller hit a tree as the airplane crash-landed in a forest. The post-crash fire had re-annealed the forged material, leading the consultant to conclude mistakenly that the propeller had been fabricated from a low-strength plain casting.

The significance of chemical environments and the importance of recognizing the critical damage mechanism are illustrated by a case involving a marine application of pretensioned high-strength bolts below the waterline [40]. A proposal was made to replace ferritic alloy bolts which had not experienced any service failures with precipitation-hardening austenitic bolts. This proposal was based solely on the fact that the austenitic alloy had a greater crack-nucleation fatigue strength, as demonstrated by rotating-bending tests in 3.5% NaCl (simulated seawater) solution. The proposal was rejected because the austenitic alloy was found to be highly susceptible to pit-corrosion cracking in stagnant seawater, particularly in harbors with significant concentrations of dissolved H_2S.

The marine bolt case also illustrates the subtle effects of counterintuitive structural behavior. Three independent teams (one led by the author) analyzed the bolt stresses based on the specified pretension conditions. A U.S. Navy structures specialist later made and analyzed strain measurements on subscale and full-scale hardware. He discovered the existence of a prying action which effectively eliminated the pretension conditions at some locations and thereby overloaded some of the bolts. The bolt material was ultimately changed to K-Monel to provide increased fatigue strength without sacrificing corrosion resistance.

Parts or materials often deviate from specifications or nominal properties. Such situations are insidious because quality-control actions may not detect the problem before many bad items have entered service. Quantitative damage analysis must be applied with careful consideration of the probable extent of degradation in material properties. Corrective action requirements vary depending on the potential for loss of structural integrity, the safety margin remaining after the deviation is accounted for, and the ability to trace the bad items. Table 3.2 summarizes several cases.

The foregoing summaries have only scratched the surface of the rich mine of possible failure causes. Damage analysis has its uses in such cases but should be subordinated to an initial approach that emphasizes clinical description. These cases also show that one must guard against forcing a problem into metallurgical or mechanical as well as damage-rate models.

3.7. Concluding remarks

The body of practices composing failure mechanics tends to expand by ad hoc response to specific incidents. There is nothing like a good crisis to stimulate interest in the otherwise mundane jobs of codifying diverse experimental results and simplifying analytical methods for practical application. However, it is useful to at least consider possible applications ahead of specific requirements so that crisis need not become panic. There follow a few suggestions of where the theoretical and experimental engineering sciences might make new contributions to failure mechanics in the near future.

More work is needed on the mechanics of three-dimensional crack propagation and fracture. It is only an assumption that growth rate and fracture toughness properties measured in tests with two-dimensional characteristics apply to three-dimensional configurations via K-correlations. For example, consider the rail head 'detail fracture' — one of several kinds of rail cracking of great concern to the U.S. railroad industry in the current environment of increasing traffic density on aging track [20]. The detail fracture grows with a small plastic zone ahead of the crack front, and is thus an ideal candidate for correlations based on linear fracture mechanics.

Figure 3.23 illustrates a typical detail fracture which developed in a rail under heavy freight service. The evident curvature of the crack surface raises an immediate question about the validity of K-correlation, which by mathematical necessity assumes plane surfaces. Figure 3.24 illustrates another detail fracture which is typical of the crack growth observed at the U.S. Transportation Test Center [41] as part of a continuing joint industry/government/university research project. This crack developed under simulated heavy freight operations with a special characteristic: about one million gross tons between traffic reversals. The ridged pattern on the fatigue part of the fracture surface suggests that detail fractures propagate under combined Mode-I/Mode-II and Mode-I/Mode-III loading.

Table 3.2. Some failures caused by deviations of material properties or processing practices.

Cause	First indication	Potential failure consequences	Traceability	Corrective action
Core shift during casting of hollow members of steel truck frames for a fleet of 750 subway cars	Visible cracks detected at details during scheduled maintenance	Loss of motor supports; dropping of motor and derailment of car	Confined to part of production run affecting about 250 cars	Bi-weekly inspection until replacement
Gag-straightening (3-point bend press) of over-cambered rails	Kinks detected on rail running surface during first year after installation in track	Slightly increased wheel/ rail impact loads; possible deepening of kinks as tonnage accumulates	Specific mill and 'year rolled' markings; track sections identified from railroad construction records	Grind running surface to re-move kinks; inspect annually to see if kinks reappear
Stacking 2024 aluminum sheet stock too close together in quench pit, causing over-aging of random areas	Aircraft manufacturer's quality control test disclosed poor strength after lot had been through major parts production	Wing skin panels unable to sustain design ultimate load	Production lot identified; parts traced to specific airplanes	Immediate replacement of all wing skin panels from affected lot
Co-drilling of built-up steel/aluminum wing spar details; steel chips carried by drill bit scored fastener hole walls in aluminum parts	Early cracking detected in full-scale airframe fatigue test	Loss of airplane by rapid fracture of undetected crack in a single-panel wing box	Critical structural areas identified	Special inspection and fastener hole rework; change fabrication procedures in manufacturing specifications
Inability to control scatter in 7178-T6 fracture toughness by means of process specification	Aircraft manufacturer's laboratory evaluation of wing skin panels on aging aircraft	Loss of airplane by rapid fracture path through areas with low local toughness	None	Wing reskin program in affected fleet; delete 7178-T6 from list of materials approved for safety-critical structure

145

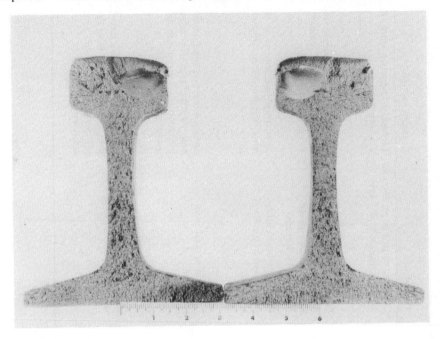

Figure 3.23.   Typical service growth pattern for a detail fracture in a rail head (the fatigue growth occurred in the smooth oval region; the remainder of the fracture surface was produced by a single static overload in the laboratory).

Mode I K-correlation is being applied to the problem of detail fracture life prediction today, but there is no guarantee that the application will succeed. Supporting research is now in progress to determine if the strain-energy density approach discussed elsewhere in this book can be reduced to application independent of the limitations of conventional fracture mechanics, which requires plane crack surfaces, assumes constant crack-front shape, and does not specify the $K_I/K_{II}/K_{III}$ interaction.

Residual stress research is another area that can benefit failure mechanics. Well established methods exist for calculating residual stresses in many kinds of welds [42]. Some measurements have been made of the residual stresses developed in rails by mill practice [4] and by service loads [43, 44], but more are needed to characterize the range of possible service conditions. An experimental foundation and good general principles exist for estimating the residual stresses produced by quenching steel-alloy round stock [4], but additional research is needed to define the service stress conditions in items such as axles. Recent limited measurements suggest that these residual stress fields may be more complex than first thought because of the quenching process itself, the effects of post-quench machining, and/or service stress relaxation [45]. On the analytical side, additional K-solutions are needed for the kinds of stress gradients associated with residual stresses, and the problem of interaction between a propagating crack and a residual stress field must be addressed.

Crack propagation in built-up structures such as airframes can be strongly influenced by load-shedding, e.g., when the crack passes a line of fasteners and causes localized fastener failures. One study has shown that this effect can actually reduce the rate of

146

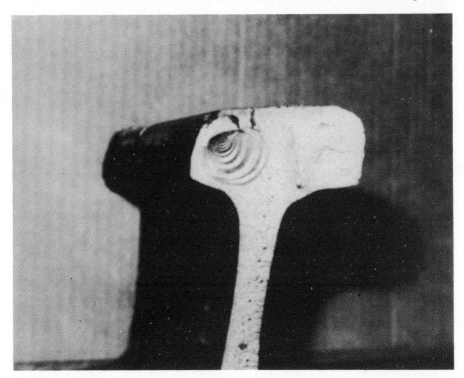

Figure 3.24. A detail fracture grown in the U.S. transportation Test Center track (the ridged appearance of the fatigue region is associated with the unique traffic pattern on the test track).

crack growth if the fasteners have sufficient flexibility, but the study was limited to specific locations in one airframe [46]. Fastener-flexibility models depend on many design details and will require parametric numerical analysis coupled with verification experiments to reduce the results to designer's-chart form for failure mechanics applications.

The designer's-chart format was one of the objectives of a recently completed research program on ductile fracture [47]. In this work a three-dimensional Dugdale-type model was applied to the analysis of surface crack growth in ductile line-pipe steels, using crack-opening and crack-tip-opening displacements as the correlating parameters. The analysis involves solution of an elastic-perfectly plastic boundary-value problem with the location of the yield boundary treated as one of the unknowns. The fundamental procedures are too complex for direct application to failure cases, but the designer's charts can be used when combined with COD/CTOD-based crack growth rate data. This is another line of research which avoids the limitations of conventional fracture mechanics. It may eventually prove useful in other applications such as low-cycle fatigue and creep-fatigue interaction in reactor hot loops or the hot sections of turbine engines.

The foregoing suggestions are certainly incomplete and probably biased, since there is no way to foretell what the next case will involve. We can be sure, however, that mechanical failures will continue to happen as long as man builds and uses structures,

147

and that some of these failures will require better methods of analysis than those we have today. One must hope, therefore, that a failure mechanics literature will arise to continue blending the theoretical and observational aspects of the engineering sciences with service-related casework.

References

[1]   Anon., *Damage Tolerant Design Handbook*, Battelle Columbus Laboratories, Columbus, OH, MCIC-HB-01 (January 1975).

[2]   Rolfe, S.T. and Barsom, J.M., *Fracture and Fatigue Control in Structures*, Prentice-Hall, Englewood Cliffs, NJ (1977).

[3]   Orringer, O. et al., 'An Assessment of the Structural Modifications of the Grumman Flxible Advanced Design Bus', DOT Transportation Systems Center, Cambridge, MA (March 5, 1982).

[4]   Horger, O.J., 'Residual Stresses', *Handbook of Experimental Stress Analysis*, edited by M. Hetenyi, Wiley, New York (1950).

[5]   Rice, R.C., Leis, B.N. and Tuttle, M.E., 'An Examination of the Influence of Residual Stresses on the Fatigue and Fracture of Railroad Rail', *Residual Stress Effects in Fatigue*, ASTM STP 776, American Society for Testing and Materials (August 1982).

[6]   Barsom, J.M., 'Fatigue Crack Growth Under Variable-Amplitude Loading in Various Bridge Steels', *Fatigue Crack Growth Under Spectrum Loads*, ASTM STP 595, American Society for Testing and Materials, pp. 217–235 (1976).

[7]   Hudson, C.M., 'A Root-Mean-Square Approach for Predicting Fatigue Crack Growth Under Random Loading', Methods and *Models for Predicting Fatigue Crack Growth Under Random Loading*, edited by J.B. Chang and C.M. Hudson, ASTM STP 748, American Society for Testing and Materials, pp. 41–52 (1981).

[8]   Hahn, G.J. and Shapiro, S.S., *Statistical Models in Engineering*, Wiley, New York (1967).

[9]   Dowling, N.E., 'Fatigue Failure Predictions for Complicated Stress-Strain Histories', *Journal of Materials*, 7(1)(1972).

[10]   Osgood, C.C., private communication.

[11]   Orringer, O., 'Analysis of Preproduction Operational Test Data for New Truck Bolster for Metroliner', Aeroelastic and Structures Research Laboratory, MIT, Cambridge, MA, ASRL TR 185-5 (November 1977).

[12]   Bendat, J.S. and Piersol, A.G., *Random Data: Analysis and Measurement Procedures*, Wiley, New York (1971).

[13]   Orringer, O., 'Amfleet Disc Brake Evaluation', DOT Transportation Systems Center, Cambridge, MA, PM-743-C-14-77 (February 1980).

[14]   Orringer, O. and Owings, R.P., 'Results from Operational Test of Hollow and Solid M-2 Axles', DOT Transportation Systems Center, Cambridge, MA, and Ensco, Inc., Springfield, VA, PM-76-C-14-1 (September 1982).

[15]   Holpp, J.E. and Landy, M.A., 'The Development of Fatigue/Crack Growth Analysis Loading Spectra', Structures Division (ASD/ENF), Wright-Patterson Air Force Base, OH, AGARD Report No. 640 (January 1976).

[16]   Dill, H.D. and Saff, C.R., 'Effect of Fighter Attack Spectrum on Crack Growth', McDonnell Aircraft Company, St. Louis, MO, AFFDL-TR-76-112 (November 1976).

[17]   McCarthy, J.F. Jr., Tiffany, C.F. and Orringer, O., 'Application of Fracture Mechanics to Decisions on Structural Modifications of Existing Aircraft Fleets', *Case Studies in Fracture Mechanics*, edited by T.P. Rich and D.J. Cartwright, U.S. Army Materials and Mechanics Research Center, Watertown, MA, AMMRC-MS-77-5 (June 1977).

[18]   Low, G.M., et al., *Improving Aircraft Safety*, National Academy of Sciences, Washington, DC (1980).

[19]   Fisher, J.W., et al., 'Detection and Repair of Fatigue Damage in Welded Highway Bridges',

Transportation Research Board, National Research Council, Washington, DC, NCHRP Report 206, (June 1979).

[20] Orringer, O. and Bush, M.W., 'Applying Modern Fracture Mechanics to Improve the Control of Rail Defects in Track', *American Railway Engineering Association Bulletin 689*, Vol. 84, pp. 19–53, (1983).

[21] Palmgren, A., 'Die Lebensdauer von Kugellagern', *ZVDI*, Vol. 68, pp. 339–341, (1924).

[22] Miner, M.A., 'Cumulative Damage in Fatigue', *Journal of Applied Mechanics*, 12, pp. A159– A164 (1945).

[23] Sines, G., 'Behavior of Metals under Complex Static and Alternating Stresses', *Metal Fatigue*, edited by G. Sines and J.L. Waisman, McGraw-Hill, New York, pp. 145–169 (1959).

[24] Gatts, R.R., 'Application of a Cumulative Damage Concept to Fatigue', ASME Paper No. 60-WA-144 (July 1960).

[25] Osgood, C.C., *Fatigue Design*, Wiley, New York (1970).

[26] Schijve, J., 'The Accumulation of Fatigue Damage in Aircraft Materials and Structures', NATO Advisory Group for Aerospace Research and Development, AGARD-AG-157 (January 1972).

[27] Waterhouse, R.B., *Fretting Corrosion*, Pergamon Press, New York (1975).

[28] Scutti, J.J., 'Fatigue Properties of Rail Steel', SB thesis, Department of Materials Science and Engineering, MIT, Cambridge, MA (June 1982).

[29] Rungta, R., Rice, R.C. and Buchheit, R.D., 'Post-Service Rail Defect Analysis', Battelle Columbus Laboratories, Columbus, OH, interim report under Contract DOT-TSC-1708 (July 1982).

[30] Forman, R.G., Kearney, V.E. and Engle, R.M., 'Numerical Analysis of Crack Propagation in Cyclically Loaded Structures', *Transactions of the ASME, Journal of Basic Engineering*, Vol. 89, pp. 459–464 (September 1967).

[31] Schijve, J., 'Observations on the Prediction of Fatigue Crack Growth Propagation Under Variable-Amplitude Loading', *Fatigue Crack Growth Under Spectrum Loads*, ASTM STP 595, American Society for Testing and Materials, pp. 3–23 (1976).

[32] Chang, J.B. and Hudson, C.M. (eds.), *Methods and Models for Predicting Fatigue Crack Growth Under Random Loading*, ASTM STP 748, American Society for Testing and Materials (1981).

[33] Willenborg, J.D., Engle, R.M. and Wood, H.A., 'A Crack Growth Retardation Model Using an Effective Stress Concept', U.S. Air Force Flight Dynamics Laboratory, Wright-Patterson Air Force Base, OH, AFFDL-TM-FBR-71-1 (January 1971).

[34] Elber, W., 'Equivalent Constant-Amplitude Concept for Crack Growth Under Spectrum Loading', *Fatigue Crack Growth Under Spectrum Loads*, ASTM STP 595, American Society for Testing and Materials, pp. 236–250 (1976).

[35] Ritchie, R.O., 'Near-Threshold Fatigue Crack Propagation in Steels', *International Metals Reviews*, Nos. 5–6, pp. 205–230 (1979).

[36] Sih, G.C., *Handbook of Stress Intensity Factors for Researchers and Engineers*, Institute of Fracture and Solid Mechanics, Lehigh University, Bethlehem, PA (1973).

[37] Shah, R.C. and Kobayashi, A.S., 'Stress Intensity Factor for an Elliptical Crack Approaching the Surface of a Plate in Bending', *Stress Analysis and Growth of Cracks, Proc. 1971 National Symposium on Fracture Mechanics, Part 1*, ASTM STP 513, American Society for Testing and Materials, pp. 3–21 (1972).

[38] Crandall, S.H. and Mark, W.D., *Random Vibrations in Mechanical Systems*, Academic Press, New York (1973).

[39] Orringer, O., 'Rapid Estimation of Spectrum Crack Growth Life Based on the Palmgren-Miner Rule', *Computers & Structures* (1984).

[40] Kerwin, J.E., Orringer, O. and Pelloux, R.M., 'Failure and Safety Analysis of Controllable-Pitch Propellers for the USS Barbey (DE 1088) and USS Spruance (DD 963)', Department of Ocean Engineering, MIT, Cambridge, MA (April 1975).

[41] Steele, R.K. and Reiff, R.P., 'Rail: Its Behavior and Relationship to Total System Wear', *Proc. 1981 FAST Engineering Conference*, Denver, CO, pp. 115–164(a)(November 1981).

[42] Masubuchi, K., *Analysis of Welded Structures*, Pergamon Press, New York (1980).

149

[43] Schilling, G.C. and Blake, G.T., 'Measurement of Triaxial Residual Stresses in Railroad Rails', U.S. Steel Corporation Research Laboratory, Monroeville, PA, Report No. 76-H-046 (019-1) (February 1980).

[44] Groom, J.J., 'Determination of Residual Stresses in Rails', Battelle Columbus Laboratories, Columbus, OH, final report under Contract DOT-TSC-1426 (August 1982).

[45] Slutter, R.G., 'Residual Stress Investigation of M-2 Axles', Fritz Engineering Laboratory, Lehigh University, Bethlehem, PA, Report No. 200.82.762.1 (September 1982).

[46] Dempster, J.B. et al., 'Ad Hoc Group Review of C/KC-135 Structure', Boeing-Wichita Company, Wichita, KA (February 1977).

[47] Erdogan, F., 'Theoretical and Experimental Study of Fracture in Pipelines Containing Circumferential Flaws', Department of Mechanical Engineering and Mechanics, Lehigh University, Bethlehem, PA, final report under Contract DOT-RC-82007 (August 1982).

Critical analysis of flaw acceptance methods

4.1. Introduction

A brief review is made on the flaw acceptance methodology. The analysis of distribution of defects and the associated nondestructive testing capability are analysed with special emphasis in aircraft structure applications and failure probability analysis. The main steps of a damage tolerance assessment program are also reviewed showing how Fracture Mechanics can be successfully applied to establish operational limits and inspection requirements.

The flaw acceptance criterion BS PD 6493 (1980) is analysed in some detail and its methodology discussed. The brittle fracture analysis for stress levels below the yield stress can give accurate and realistic results. However, the elastic-plastic method is too conservative due to lack of representative experimental toughness data and more accurate computations of stress and strain fields near stress concentration areas. The simplified fatigue assessment method needs to be checked against appropriate experimental and theoretical data before it can be used as a flaw analysis method. Otherwise its application will have little practical interest due to the high degree of safety introduced.

4.2. Defects: distribution and non-destructive testing capability

All structures contain flaws, namely metallurgical defects such as inclusions or porosity, or microscopic crack-like flaws, either intrinsic or initiated in service. The former are those designed in the structure such as windows in an aircraft. The latter are usually due to the occurrence of stress concentrations features in many components in the form of rapid changes in section, keyways, splines or other forms of notches including the stress concentrations in welded, riveted or bolted joints. These are favourable sites for the initiation of cracks which may arise due to cyclically fluctuating loads (fatigue), corrosion or stress corrosion cracking. Table 4.1 taken from [1] shows a comprehensive list of typical material defects.

It is now widely accepted that Fracture Mechanics can be applied either to evaluate the severity of a known defect or to establish tolerable flaw acceptance levels. Flaw

Table 4.1.   Typical material defects.

Defects existing in mill products	Defects produced in service
Chemical contamination	*Mechanical damage*
Inclusions, dirt	Particle damage, tool marks
Segregation	Improper repair, maintenance
Laminations	Fatigue cracks
Internal defects	Fretting
Porosity	Creep
Pipes	*Environmental damage*
Cracks	Corrosion
Surface defects	Stress corrosion, corrosion fatigue
Cracks, tears	Bacterial degradation
Laps, pits	Thermal degradation
Scratches	
Distortion	

Defects produced by processing	
Metal removal	*Heat treatment*
Cracks	Cracks
Tool marks, gauges	Distortion
Metal finishing	Decarburization
Cracks in coating, base metal	Incomplete transformations
Pits, blisters	*Joining*
Lack of adhesion, insufficient thickness	Weld defects-cracks
Hydrogen embrittlement	Incomplete fusion, residual stress
Surface contamination	Fasteners-tears, galvanic corrosion

growth analysis is now generally used as a design and quality control procedure in several high technology areas like the aircraft industry. This design approach is known as the *fail safe* approach and in order to be applied requires the knowledge, among other factors, of the size and distribution of defects in the structure. A probabilistic analysis of flaw distribution is necessary in order to assess the size of the largest initial flaw in the structure. This is a difficult and time-consuming task since it requires a considerable amount of data usually obtained by different processes, viz.:

(1) analysis of nondestructive records of flaw shape, distribution and location obtained in the actual or in a similar structure;

(2) analysis of crack propagation tests carried out in selected parts of the structure.

The latter process will be explained in more detail in the following section. The main objective of this analysis is to define the probability of finding a crack in a particular location.

Thus, a probability density function for defect size and location can be obtained by this process and related with the probability density function for tolerable defect size. This will be the maximum size of defects accepted in a certain location in order to avoid fracture (brittle fracture, fatigue, stress corrosion).

The failure probability can therefore be assessed by the evaluation of an expression of the following form

$$\text{Failure fraction} = \int n(x)\mathrm{d}x \int g(a)\mathrm{d}a \qquad (4.1)$$

where $n(x)$ and $g(a)$ are the probability density functions for defect size and tolerable defect size respectively. The $n(x)$ function can be obtained as explained in (1) and (2). The $g(a)$ function has to be estimated from fracture toughness, subcritical crack propagation data and design data (stress level, load spectrum, component geometry and location, type of flaw etc.) and is entirely obtained applying Fracture Mechanics methods.

The probability density function is influenced by many different parameters. The crack size is one of these parameters as it can be seen in Figure 4.1 which is a typical schematic curve relating the number of flaws present and detected with the crack size. The curve shows that the number of flaws detected are always less than the number of flaws present and there is a crack size a* below which the detection is not possible (Figure 4.1). This value of crack size a* is the smallest flaw that can escape detection during the inspection procedure and one can always assume with a probability of say 100% that the structure contains cracks with sizes smaller than a*. Initial cracks with

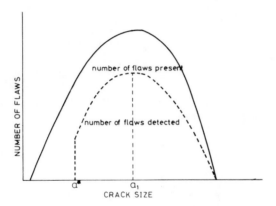

Figure 4.1. Schematic curve presenting a comparison of flaw size distribution with non-destructive testing detection limits [2].

sizes greater than a* are possible to detect and the probability of detection increases with the crack size for the small defects. Hence Figure 4.1 indicates that a starting point for the application of Fracture Mechanics is the definition with a given statistical basis of the largest initial flaw that could reasonably be present in the structure. Both the flaw-size distribution and the nondestructive testing detection curve increase with crack size till a certain crack size a_1 is reached. If a crack size is bigger than a_1 its density decreases and the corresponding number of flaws detected also decreases. There will typically be a large quantity of small defects with a decreasing number of larger ones (Figure 4.1).

The flaw size detection limit depends of many different factors including the experience of the NDT operator. Some of these factors are:

(1) material characteristics (grain size, amount and distribution of second phase particles, and other metallurgical parameters);
(2) type of flaw (planar defect, nonplanar defect, surface defect, embedded defect);
(3) processing conditions (residual stresses, surface finish, previous processing);

(4) human factors (inspector experience, personal interpretation of results, level of training);
(5) equipment or procedure (calibration of equipment, inspection procedures, sequencing of operations);
(6) detection limit criteria (required level of probability and detection limit).

Also the NDT method and all its related aspects (staff and inspection environment) must simulate those service conditions found in production and (or) service. Table 4.2, taken from a survey of the U.S. aircraft industry, indicates that NDT detection limits based on laboratory data are often completely nonrepresentative of actual service

Table 4.2.   Estimated variations in flaw detection limits by type of inspection (in mm) (in Pettit & Krupp, 1974).

NDT Technique	Surface cracks		Internal flaws	
	Processing	Fatigue	Voids	Cracks
Test specimens, laboratory inspection				
Visual[†]	1.25	0.75	+	+
Ultrasonic	0.12	0.12	0.35	2.0
Magnetic particle	0.75	0.75	7.5	7.5
Penetrant	0.25	0.5	+	+
Radiography	0.5	0.5	0.25	0.75
Eddy current	0.25	0.25	+	+
Production parts, production inspection				
Visual	2.5	6.0	+	+
Ultrasonic	3.0	3.0	5.0	3.0
Magnetic particle	2.5	4.0	+	+
Penetrant	1.5	1.5	+	+
Radiography	5.0	*	1.25	*
Eddy current	2.5	5.0	+	+
Cleaned structures, service inspection				
Visual	6.0	12.0	+	+
Ultrasonic	5.0	5.0	4.0	5.0
Magnetic particle	6.0	10.0	+	+
Penetrant	1.25	1.25	+	+
Radiography	12.0	*	4.0	*
Eddy current	5.0	6.0	+	+

[+]Not applicable. [†]Use with magnifer. [*]Not possible for tight cracks. (Based on 25-mm ferritic steel, surface 63 RMS.)

inspections. This data only provides a comparative indication of the detection limits obtained by the different techniques. As pointed out before a crack with a certain size can only be detected with a certain probability. Thus Figure 4.2 shows that the ultrasonic technique may detect a 4-mm defect in 50% of cases but a 7-mm defect can be detected almost every time. The graph also proves that the dye penetrant technique is less accurate for sizing defects than the ultrasonics for crack lengths less than about 5 mm.

As pointed out before, it is evident that defect sizing is of particular importance to assess structural integrity. The general procedure explained previously can be applied,

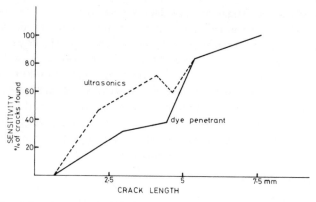

Figure 4.2. Flaw detection capability for ultrasonic and dye penetrant (Packman 1968).

for example, to define the acceptable size of defect for a given probability of failure. Two methods are used in defect analysis: the *damage tolerance assessment* and the *flaw acceptance criteria*. The damage tolerance assessment is now extensively used in the aircraft industry and it is not a different method from the flaw acceptance criteria. They only differ in the range of applicability, objectives and sequencing of analysis. In the United Kingdom a public document has been issued in 1980 [3] dealing specifically with methods for the derivation of acceptance levels for defects in fusion welded joints. Similar criteria are being developed in the same country for specific welded structures such as pressure vessels, cranes and offshore structures. In the following sections a brief description of the damage tolerance procedure and an analysis and review of the PD 6493 will be made.

4.3. Damage tolerance assessment

A damage tolerance assessment program is usually carried out either in the prototype stage of aircraft development or in the early stages of the aircraft life. The main objective of the damage tolerance assessment is to define structural life, operational limits and establish improved structural inspection requirements. An operational life in terms of flight hours is previously defined and the structural life of the aircraft, determined using Fracture Mechanics, should be greater than the operational life. A damage tolerance assessment program is normally carried out in five stages:

(1) Preliminary Damage Tolerance Assessment;
(2) Initial Quality Assessment;
(3) Stress Spectra Development;
(4) Operational Limits and Inspection Requirements;
(5) Individual Aircraft Tracking Program.

Each one of these stages will now be explained in more detail.

In the Preliminary Damage Tolerance Assessment one selects and evaluates potentially critical structural areas in the aircraft structure which could potentially affect safety of flight. Candidate critical areas are systematically identified and screened

155

using available stress and fatigue analysis data, service experience and nondestructive inspection data. Hence, the potentially critical structural items are identified, and to each location an assessment is made of its *degree of criticality*. The degree of criticality depends on several factors. For example, there are areas where a single failure would result in the loss of aircraft. These would be considered more critical than areas which could sustain a failure without loss of aircraft. Another important factor is the inspectability. Areas more difficult to inspect are considered more critical than others. In practice it means that, due to economic reasons, these areas will be inspected less frequently and this must be taken into account in the inspection intervals. Additional considerations affecting the degree of criticality are:

— material type and thickness
— strength and fatigue design margins
— crack growth rates
— stress concentrations
— load transfer
— service experience
— test data.

Usually, after a detailed analysis has been carried out, only a few areas are judged sufficiently critical to warrant detailed analysis.

The purpose of the *Initial Quality Assessment* is to assess the initial manufacturing quality of the structure and to determine the maximum size of initial flaws that could be expected. A statistical distribution of initial manufacturing flaws is obtained using the processes (1) and (or)(2) referred to in the previous section. The nondestructive record analysis is comparatively straightforward to carry out and only requires a well qualified and trained operator and an extensive collection of records for the selected critical areas. However, NDT are not able to detect flaws with sizes below the values quoted in Table 4.2. As pointed out before, one can always start the Fracture Mechanics analysis with the assumption that an initial flaw (with a size equivalent to the flaw detection limits) exists in the critical location. However, this approach may not be adequate, due to all the factors affecting the detection limits. A better method (though more expensive) is to obtain the sizes of the initial manufacturing flaws as outlined in (2). For this method the procedure is to fatigue test the selected component using a generally modified constant amplitude spectrum to allow adequate fatigue marking. The crack then initiates and propagates to failure. A fractographic analysis of the crack surfaces can reveal the initial size of the crack. After the initial sizes are obtained in each location, the data is statistically evaluated. Often a log normal probability distribution is obtained and in such a case the plot schematically presented in Figure 4.3 shows the percentage of cumulative number of flaws against its equivalent initial flaw size. Assuming a certain probability of occurrence and a given confidence limit the maximum initial manufacturing flaw size can be obtained. This value defines the initial quality assessment in the aircraft structure.

The *Stress Spectra Development* task consists of the development of baseline representative stress spectra to be used to obtain the operational limits, and inspection requirements for the critical items identified in the Preliminary Damage Tolerance Assessment. The flight-by-flight stress spectra should be representative of load

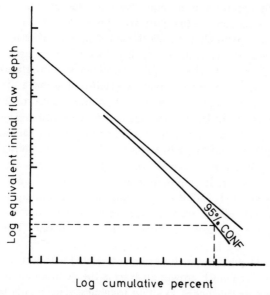

Figure 4.3. Typical statistical distribution of initial manufacturing flaws.

conditions resulting from combinations of weight, altitude, airspeed and aircraft response. The principal stresses are computed for the critical items and for each loading condition. The accuracy of the analytical stress equations can be checked by comparison with flight measured stresses.

The *Operational Limits and Inspection Requirements* are established conducting crack growth analysis and tests for the critical areas selected in the Preliminary Damage Tolerance Assessment using the results of the statistical distribution of initial manufacturing flaws and the flight-by-flight stress spectra as the stress input. The operational limits are based upon conservative assumptions to prevent loss of the worst case aircraft. The operational limit is defined as the service usage interval beyond which a potential failure of the unrepaired structure could result in loss of aircraft. The operational limits for each critical item or location can be calculated integrating an appropriate fatigue crack growth rate equation from an initial flaw size to a critical crack size. The initial flaw size could be either the initial manufacturing flaw size defined in Figure 4.3 or the largest flaw with a certain shape that would remain undetected during airframe fabrication. The critical crack sizes are dependent on the residual strength of the structure which may be defined on the basis of two requirements:

(a) the structure must be capable of sustaining design limit load;
(b) the structure must be capable of sustaining the maximum load expected during the inspection period.

The inspection period may be defined, for example, as one-half the operational limit.

The crack growth rate data selected for the analysis should be representative of the material, thickness, environment and stress ratio. The stress intensity factor formulation should be appropriate for the type of defect and loading mode. The crack integration model should account for variable amplitude loading usually occurring in realistic flight-by-flight spectra and account also for crack growth retardation. The Wheeler model [4] may be used for that purpose. This model uses the load interaction or plastic zone size concept to characterize the crack tip residual stress state created by prior spectrum loads. If the currently applied load develops a plastic zone to or past one previously developed (greatest prior elastic-plastic interface), the growth increment associated with the current load is calculated using a steady-state (no retardation) growth rate equation. Conversely, retardation is assumed if the current load develops a plastic zone smaller than the one which preceeded it. The crack growth can therefore be predicted based on accumulation of increments caused by each load application defined in the flight-by-flight stress spectra and considering crack growth retardation as explained before. Several computer routines are available for this purpose [5] which use constant amplitude crack growth rate da/dN for each load application. The Forman [6] and Paris [7] crack growth rate equations are commonly used. The Wheeler model accounts for retardation by operating directly on da/dN and reducing the constant amplitude crack growth rate. The crack growth rate under spectrum loading is computed using a shaping parameter, m, which is a data fitting parameter allowing the analysis to be correlated with test results [4].

The final element of the crack growth analysis is the determination of the critical stress intensity or fracture toughness. Fracture toughness data should be obtained for the materials, thickness and temperature of interest. Finally, the crack length increments can be computed with the application of the stress cycles defined in the stress spectra and taking into account the details mentioned before. The calculation ends when the current crack length and minimum required residual strength load produced a stress intensity equal to the fracture toughness. The number of cycles obtained in the calculation constitute the operational limit of the critical location being considered.

The corresponding crack sizes to the inspection interval chosen can be calculated using the same procedure. This calculated crack size can then be compared with the size measured during the inspection, and therefore an assessment of component or structure integrity can be done during the inspection period. If, for example, the measured crack size in the location is greater than the calculated or expected crack size the component must be removed or repaired.

The *Individual Aircraft Tracking Program* is applicable to monitor damage in individual aircraft. This turns out to be necessary because the operational limits and inspection intervals are usually established for *baseline* usage and this varies from aircraft to aircraft. More details of this method may be found in [8].

4.4. Flaw acceptance criteria

The BS document PD 6493 (1980) is probably the only official work available where a systematic procedure was developed for the derivation of defect acceptance levels. PD 6493 is applicable for welded defects which is not a very common type of defect in aircraft structures. However the method of analysis in some conditions can also be applied to derive flaw acceptance levels in aircraft structures.

PD 6493 (1980) uses the Engineering Critical Assessment design philosophy. This is a compromise between the traditional safe life approach and the fail safe approach. Thus the quality control standards are based in traditional engineering practice defined in the codes currently in use in welded constructions. The design is therefore a safe life design, crack initiation is not tolerated and periodic inspection requirements are not necessary. However, if a flaw is detected in the structure, a critical assessment should be carried out taking into account the type of structure and its use. Therefore a decision may be taken on whether a flaw should be repaired or not, making it possible to avoid expensive weld repairs which would have always been necessary in the case of planar type flaws, when applying the current codes of construction. Moreover, repair of welds can alter the material behaviour in the weld metal and heat-affected zone, and changes in toughness are to be expected. It is known that repair welding, although frequently inefficient, will reduce the number of flaws but it may also introduce regions of lower metallurgical quality (lower toughness). Another important factor in repair welding is that the tolerable defect size will also be affected by the consistency of the metallurgical quality of the weldment.

In the BS document the flaws are classified into two groups: planar defects and nonplanar defects. The planar defects are assessed using Fracture Mechanics methods while for the nonplanar the traditional Charpy-V and $S-N$ approach are used. The main modes of failure are identified and a detailed assessment is explained for brittle fracture and fatigue failure.

Assessment for brittle fracture

Brittle fracture may occur as a result of low temperatures, stress concentrations, high values of thickness and strain rate, metallurgical effects caused by the material micro-structure and metallurgical alterations due to the welding process. If a flaw is present, unstable fracture occurs when the stress intensity factor reaches the appropriate fracture toughness value for the material, thickness and temperature. Fracture toughness may be expressed in terms of the stress intensity factor K_{1c} or K_c and, where appropriate, in terms of a critical crack opening displacement COD, δ_c or J integral J_{1c} or J_c. For example, in welded joints, fracture toughness should be obtained in the parent material, weld metal and heat-affected zone in specimens tested with the same material, thickness, temperature and environment as in service. The fracture toughness values should preferably be estimated from test results using currently available specifications (e.g., ASTM E 399-74, BS 5447).

The flaw assessment requires a set of data, namely:

(1) position and orientation of defect;
(2) structural and weld geometry;
(3) stresses and temperatures including transients;
(4) conventional yield or proof stress in tension and Young's modulus;
(5) fracture toughness data;
(6) bulk corrosion and stress corrosion cracking (K_{1scc}) data.

The position and orientation of defect can be obtained using NDT techniques following the appropriate specifications. Suitable allowances should be incorporated in the assessment of defect sizes to cover intrinsic and measurement errors involved and thereby ensure conservative assessment of defect severity. The stresses to be considered in the analysis are all the calculated or measured stresses in the defect zone without taking into account the stress intensity magnifying factor due to the defect itself. In welded joints a residual stress value equal to the yield stress of the material should be considered. In stress-relieved structures the residual stress may not be zero and an estimate of the actual value should be made. For planar defects only the stress component normal to the plane of defect should be used since the assessment is predominantly for Mode I loading.

The dimensions of the planar defects are defined by the dimensions of their containment rectangles as shown in Figure 4.4(a)–(c). An interaction criteria is defined for multiple arrays of defects in different positions [3]. If the defects are sufficiently close for interaction to be possible they should be treated as a single defect with an

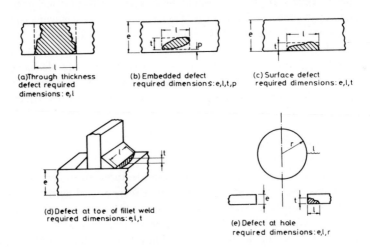

(a)Through thickness
defect required
dimensions: e,l

(b) Embedded defect
required dimensions:e,l,t,p

(c) Surface defect
required dimensions: e,l,t

(d)Defect at toe of fillet weld
required dimensions:e,l,t

(e) Defect at hole
required dimensions:e,l,r

Figure 4.4.   Defect dimensions defined in PD 6493: 1980.

effective dimension. If the defect size (t_x, l_x) is such that a through thickness defect of this length would cause yielding of the remaining cross section or local collapse, a recategorization procedure should be carried out in order to evaluate whether a surface or embedded defect should be recategorized as a through thickness defect or a surface or through thickness defect respectively with a characteristic size.

The BS document [3] makes a distinction between situations where the stress levels are below the yield stress with a valid K_{1c} for the material, and those cases where the stress levels exceed the yield stress or K_{1c} is not valid. In the first case, the stress intensity factor should be obtained using LEFM considering the appropriate location and geometry of the defect. Several equations are presented in [3] applicable to the more common type of defects found in welded joints. Alternatively, other stress intensity factor solutions may be found in [9] and [10]. If the solution is not available in the literature, K may be obtained by applying numerical or experimental methods described in the literature [11]. The defect will be accepted if the calculated stress intensity factor is $\leqslant 0.7 K_{1c}$. If a valid K_{1c} is not obtained the document recommends the use of a critical COD value applying the procedure for stress values above the yield stress. The acceptance criterion $K_I \leqslant 0.7 K_{1c}$ was arbitrarily chosen giving a factor of safety of 2 on defect size and 1.4 on stress and stress intensity for defects lying in a uniform stress field. For some part thickness defects and long surface defects the safety factor may fall to about 1.25 due to the rapid increase of stress intensity factor with crack depth.

In the second case an 'effective' defect parameter $\bar{a}$ is determined and compared with a tolerable defect parameter $\bar{a}_m$. The value of $\bar{a}$ depends on the actual defect geometry and the applied stress level and its values were derived from modified linear elastic solutions assuming a constant strain level at the peak value, effective over the full defect length. If steep gradients of stress and strain are present near the defect zone this will inevitably give conservative estimates of allowable defect sizes. For a total stress level in the defect zone less than two times the yield stress σ_{ys}, the tolerable defect parameter is given by the equation

$$\bar{a}_m = C(\delta_c/\epsilon_y) \tag{4.2}$$

where C is a constant given in Figure 4.5, ϵ_y is the yield strain (σ_{ys}/E) and δ_c is the so called 'critical value of COD'. The constant C is presented as a 'design curve' function of the ratio principal applied stress to yield stress. For long through thickness defects in curved shells or defects recategorized as such which are subjected to pressure loading, a further correction factor is defined in the document for the value of $\bar{a}_m$. When the stress level excluding the residual stress is greater than $2\sigma_{ys}$ the document recommends a full elastic-plastic stress analysis to determine the maximum equivalent plastic strain levels which would occur in the region containing the defect if the defect were not present. Some procedures are suggested as an estimation method.

The treatment for stresses above yield stress is based upon an applied strain/yield strain criterion. It is known that when the sum of the stresses does not exceed twice

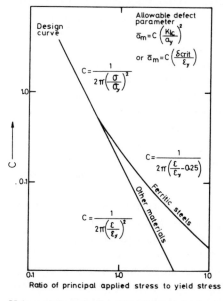

Figure 4.5. Values of the constant C for different loading conditions [3].

the yield stress direct conversion of stress to strain ($\sigma = E/\epsilon$) is acceptable. Hence the abcissa of Figure 4.5 is given as the ratio of applied stress to yield stress. However, when these stress values exceed $2\sigma_{ys}$, plastic strains may occur which are in excess of the equivalent elastic value. In these cases, and as pointed out before, an estimate of the equivalent applied strain should be made and the abcissa of Figure 4.5 taken as the ratio of applied strain to yield strain.

The approach described for the acceptance criterion for stress levels above the yield stress is no doubt too conservative due to four main reasons:

(1) The assumption of a constant strain level at the peak value effective over the full defect length is conservative for the estimation of both allowable defect sizes $\bar{a}$ and tolerable defect sizes $\bar{a}_m$.

(2) Even with the estimation of the equivalent plastic strain when $\sigma > 2\sigma_{ys}$ the results will be conservative because the same 'design curve' is applicable. Anyway, this situation will be of little practical interest since very few materials can tolerate local stresses twice the value of yield stress without fracture.

(3) The 'design curve' has a safety factor of 2 incorporated to ensure that the allowable defect size is smaller than the critical size. This can also lead to very conservative results, especially for higher values of stress.

(4) Critical COD values δ_c used in equation (4.2) are known to vary considerably with test piece geometry, and crack size. If the upper bound values are used, again very conservative estimates can be obtained. This conservatism is also considered in the 'design curve' based on the analysis of results of wide plate fracture tests, compared with minimum values from COD tests obtained in specimens with different size and geometry.

The procedure discussed so far only applies to flaws with known location and geometry. It is also possible in the same document to estimate maximum acceptable flaw sizes for different types of defect and locations. As pointed out before, the locations should be carefully considered and the assessment should be made for three main defect categories (surface, embedded and through-the-thickness defects). It is necessary to know the stress levels with reasonable accuracy and, for welded joints, fracture toughness should be evaluated in the parent plate, weld metal and heat affected zone. It is also convenient to carry out a statistical analysis of stress and material properties variation to allow a probabilistic determination of the tolerable defect size (function $g(a)$ in equation (4.1). A comparison can then be made between the computed maximum acceptable flaw sizes and the nondestructive detection limits. The analysis can therefore predict a failure probability.

As a final comment it should be mentioned that PD 6493 (1980) methods are adequate for the assessment of brittle fracture under nominally linear elastic conditions (stress levels below the yield stress and with valid K_{1c} data for the situation of interest). The approach for stress levels above the yield stress is justifiably too conservative and may lead to unrealistic estimations of flaw acceptance levels. More work is necessary in this field to improve the theoretical estimates and the fracture toughness measurements in the elastic-plastic regime.

Assessment for fatigue

Fatigue failure initiated in stress concentration areas is usually described as being mainly a crack propagation phenomenon occupying the total of the fatigue life. This approach is widely used in the fatigue assessment of welded joints where the fatigue crack initiates due to the stress concentration of the weld. Hence the crack initiation phase is not considered in the analysis because it is assumed that fatigue crack propagation takes place immediately from a preexisting weld defect.

PD 6493 (1980) describes two methods for the fatigue assessment of both planar and surface non-planar defects. The first method is the general fatigue crack propagation method using the appropriate fatigue crack propagation equation. The predicted number of cycles to failure is obtained by integrating this equation from an initial known flaw size to a critical crack size for unstable fracture. An accurate stress intensity factor solution must be known for the defect and geometry and, as emphasized before, the cyclic crack growth equation should be obtained in very similar conditions to those obtained in practice (same material, thickness, mean stress, environment, temperature, etc.). Provided the stress intensity formulation is available the analysis is comparatively straightforward for constant amplitude loading. For variable amplitude loading a linear damage law is assumed with nò interaction effects. Since this method requires advanced research and development facilities, which may not be available for most of the users, the document describes a simplified method not requiring experimental data for the analysis.

The simplified method assumes the Paris' crack propagation law [7] with a value of $m = 4$ for the exponent valid for all structural steels. This value of m is the reciprocal of the assumed exponent[*] of the $S-N$ curves obtained in constant amplitude loading of welded joints as predicted in the BS 5400 code [12] for metallic bridges and other similar structures. The value of C was adjusted to cover most of the experimental data obtained in welded steels and both values of m and C are claimed to be conservative. Figure 4.6 shows the design $S-N$ curves for welded steel and aluminium alloys. Similar curves are available for stress-relieved steels. Each curve corresponds to a certain quality category ranging from $Q1$ to $Q10$ where $Q1$ is the higher grade and $Q10$ the lower grade. The quality categories are related with initial flaw sizes not specified in the document. Thus $Q1$ curve is for an initial flaw size smaller than $Q2$, etc. For variable amplitude loading an equivalent constant amplitude stress range may be computed and used in Figure 4.6. The first six of these $S-N$ curves are identical (for steels) to those appropriate to the 97.5% survival limit for the joint classes named C, D, E, F, $F2$ and G in [12]. If the constant amplitude stress range or its

---

[*]The Paris' equation $da/dN = C(\Delta K)^m$ may be transformed in an equivalent $S-N$ curve given by the equation

$$\sigma^m N_f = \frac{I}{C}$$

where

$$I = \int_{a_0}^{a_f} \frac{da}{(Y\sqrt{\pi a})^m}$$

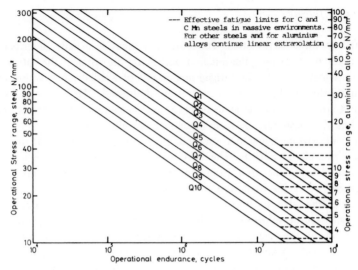

Figure 4.6.   Stress/endurance relationships function of quality categories for a welded steel and aluminium alloys [3].

equivalent is known together with the predicted or expected number of cycles the designer can get a point in the diagram of Figure 4.6 and therefore define the quality category.

The assessment of known defects may be carried out in a graphical form for both 97.5 and 99.5% survival probability. Several graphs are presented for the most important types of defect including defects at the fillet of weld toes. Surface and embedded defects are converted previously to equivalent 'long' surface and embedded defects with a straight crack front. The critical crack size for unstable fracture may be computed from critical toughness data or can be assumed to be the thickness or other significative structural dimension to which fatigue growth is permitted. A stress parameter S is computed and a quality category ranging also from $Q1$ to $Q10$ is defined for some minimum values of this parameter. The defect is acceptable if the actual quality category obtained by the analysis is higher or the same as the quality category required by the design and defined in Figure 4.6.

Both methods are also applicable to estimate tolerable flaw sizes. The calculation is carried out in reverse order starting with the final crack size and working backwards to obtain the initial flaw size for the accounted number of design fatigue cycles. The defect size at this stage is the tolerable initial flaw size.

So, basically what the simplified method does is to compare an experimental $S–N$ curve for a plain welded joint (joint shown in Figure 4.6) with a theoretically derived crack propagation $S–N$ curve obtained from computational results such as those presented in Figure 4.7. The method can therefore lead to unrealistic and conservative results since some important parameters were not taken into account like crack initiation and thickness. In an attempt to overcome these problems conservative values of some variables were taken which may lead to unrealistic results. For example Figure 4.7 presents the computational results for the stress parameter S as a function of $\bar{a}_0$ or $\bar{a}_m$ for different plate thicknesses and for a long defect at the toe of a butt

164

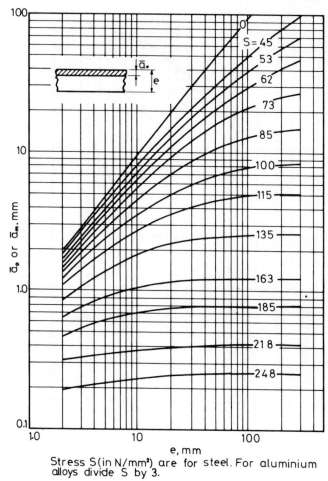

Figure 4.7. Theoretical estimates of $\bar{a}_0$ and $\bar{a}_m$ against thickness for different values of the stress parameter S. Surface defects at the toe of a butt weld in a flat plate in tension [3].

weld in tension. These results and other similar plots obtained for defects in fillet weld joints have not yet been checked against experimental data obtained by fatigue testing of components with different thicknesses, initial flaw sizes and locations.

Moreover, recent theoretical stress intensity factor studies using the finite element method have shown that fatigue crack propagation in fillet weld joints depends on other geometrical variables of the joint, like the thickness of the attachment plate, weld angle and leg length [13], as it can be seen in Figure 4.8 where the tolerable initial defect depth is plotted against the plate thickness for different attachment thickness and weld size. Another significant parameter, especially for long life prediction, is the crack initiation phase. The author has recently found that crack initiation in fillet welded joints of tubular sections in bending [14] may contribute for more than 50% of the fatigue life for endurances above 10^5 cycles. A fatigue life prediction method in welded joints must consider the crack initiation period. Comparison of $S-N$ curves obtained from basically defect-free joints with equivalent theoretical crack propagation

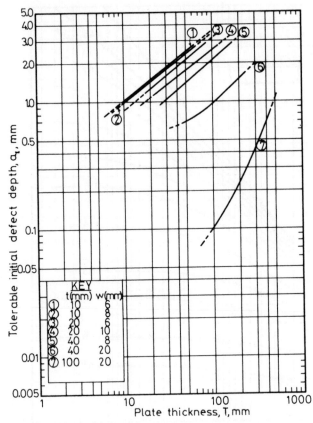

Figure 4.8.   Effect of absolute attachment thickness and weld size on initial defect sizes for fillet welded joints under fatigue loading [13].

results is only valid if the crack initiation phase is taken into account and deducted from the total fatigue life. Also the crack propagation model of the joint should be identical to the S–N curve of the joint.

The simplified method described in PD 6493 (1980) should not be used unless appropriate crack initiation data* is available for the design S–N curves and the theoretical defect acceptance curves (Figures 4.7 and 4.8) are confirmed by some experimental data. These curves should be obtained with the same joint geometry, loading conditions, flaw type and location.

So far as aircraft structures are concerned and despite the fact that, as pointed out before, welded joints are not usually found, the same basic analysis is applicable. The crack still initiates in a stress concentration area usually located near a fastener or a hole. Data of the same type as shown in Figure 4.7 and 4.8 should be obtained and conveniently checked with experimental results.

---

*Crack initiation data should cover the number of cycles from the beginning of the test to a certain minimum initial flaw depth.

4.5. Conclusions

The flaw acceptance methodology was reviewed with special emphasis in applications related with aircraft structures. The importance of an adequate sizing of defects was outlined. It has been shown that a probabilistic analysis of flaw distribution is necessary to obtain the failure probability in the structure. Flaw distribution in its turn depends on NDT capability and detection limits, and a brief presentation was given of the testing performance and accuracy of the main NDT techniques.

The application of flaw acceptance methods to fatigue is particularly important in aircraft structures and constitutes the damage tolerance assessment program extensively used in the aircraft industry. A brief description was given of the main stages of this program and also its methodology was discussed. It was seen that fracture mechanics methods are applied to define operational limits and inspection requirements as part of the damage tolerance assessment.

The BS document PD 6493 (1980) is an example of a flaw acceptance criterion specifically derived for defects in fusion welded joints which applies fracture mechanics both for brittle fracture and fatigue. Its methodology was discussed since it is also relevant to failure analysis in aircraft structures. The author believes that the method put forward in PD 6493 (1980) can be applied successfully for brittle fracture where the stress levels are below the yield stress. The elastic plastic analysis (also called brittle fracture) is very conservative due to all uncertainties associated with the COD concept and its experimental measurement. The method can be significantly improved by a more accurate computation of the elastic-plastic stress and strain field near notches and other stress concentration areas where the crack usually initiates. It is clear that a single 'design curve' as presented in PD 6493 (1980) may not be adequate to describe elastic-plastic crack growth. Instead, a compendium of solutions of appropriate elastic-plastic fracture parameters (COD, J integral, strain energy density, etc.) should become available for the more important practical cases. The theoretical estimations should be related with fracture toughness values obtained in similar geometries and loading conditions.

The simplified fatigue assessment method for flaw acceptance presented in PD 6493 (1980) is also of little practical interest due to its inherent high degree of safety. Crack initiation is not taken into account in the design $S-N$ curves used for comparative purposes and the theoretical crack growth curves do not consider the influence of important geometrical variables. Much more work is necessary to compare theoretical estimations of fatigue crack growth from an initial defect size with experimental data obtained in similar conditions. These points should also be considered in any fatigue assessment of defects in aircraft structures.

References

[1] Garrett, G.G., 'Fracture Mechanics and the Assessment of Structural Reliability'. In: *Engineering Applications of Fracture Mechanics*, Pergamon Press, London (1980).

[2] Pettit, D.E. and Krupp, W.E., 'The Role of Nondestructive Inspection in Fracture Mechanics Application'. In: *ASM 3 Fracture Prevention and Control*. American Society of Metals (1974).

[3] BS: PD 6493, 'Guidance on Some Methods for the Derivation of Acceptance Levels for Defects in Fusion Welded Joints' (1980).

[4] Wheeler, D.E., 'Crack Growth under Spectrum Loading', Report No. FZM-5602, General Dynamics Corp., Ft. Worth Texas (1970).

[5] Szamossi, M., 'Crack Propagation Analysis', Report No. NA-72-94. North American Rockwell, Los Angeles, USA (1972).

[6] Forman, R.G., Kearney, V.E. and Engle, R.M., 'Numerical Analysis of Crack Propagation in Cyclic Loaded Structures', *Journal of Basic Engineering*, Trans. ASME, 89, pp. 459–464 (1967).

[7] Paris, P.C. and Erdogan, F., 'A Critical Analysis of Crack Propagation Laws', *Journal of Basic Engineering*, 85D, 9, pp. 528–534 (1963).

[8] Gray, T.D., 'Individual Aircraft Tracking Methods for Fighter Aircraft Utilizing Counting Accelerometer Data', Technical Memorandum TM-78-1-FBE, Air Force Flight Dynamics Laboratory, Dayton, Ohio, USA (1978).

[9] Tada, H., Paris, P.C. and Irwin, G., *The Stress Analysis of Cracks Handbook*, Del Research Corporation, Pennsylvania, USA (1973).

[10] Rooke, D.P. and Cartwright, D.G., 'Compendium of Stress Intensity Factors', HMSO England (1976).

[11] Sih, G.C., *Methods of Analysis and Solutions of Crack Problems*, Noordhoff International Publishing, Leyden, The Netherlands (1973).

[12] BS 5400, 'Steel, Concrete and Composite Bridges. Part 10 – Code of Practice for Fatigue' (1980).

[13] Burdekin, F.M., 'Practical Aspects of Fracture Mechanics in Engineering Design', 17th John Player Lecture, *I. Mech. Eng. Proceedings 1981* (12)(1981).

[14] Branco, C.M. and Fernandes, A.A., 'Fatigue Behaviour of Steel Rectangular Hollow Sections used in Bus Structures', presented in the International Conference on Fracture Mechanics Technology Applied to Material Evaluation and Structure Design, Melbourne, Australia, 10–13 August 1982.

Reliability in probabilistic design

5.1. Introduction

The actual load and load-carrying capability are considered as random independent variables. Attention is focused on the structural integrity of a system at different stages of loading with emphases given to identifying the critical parts and their degree of criticality. The emphasis on material and specimen testing in recent years has overshadowed the safe life approach. The requirements for safety and durability are being considered independently. Treated in this work are the main steps to assure a correct design, including the concept of factor of safety, reliability and confidence level established from test data on distribution curves.

A structure or a simple product, such as an airplane or electric bulb, should satisfy certain manufacturing requirements with regard to its performance standards (e.g., life expectancy). For instance, the total number of hours and/or number of times for the bulb to go on and off. For an airplane, the number of hours of flight and conditions of periodical maintenance are important. This type of information applies to any other product, e.g., machine tools, locomotives, buses or moving cranes.

It is well known that the *actual load* L_a to which a system or a mechanical element is subjected is a variable independent from its *load carrying capability* L_c. The airplane structure may be taken as an example:

(1) the actual load depends on the weather conditions, on the flying habits of the individual pilots, on unexpected overloads, etc.;
(2) the load-carrying capability is a function of the shape and geometry of the structural elements (rolling tolerances), of the characteristics and nature of flaws in the material imperfections introduced during the manufacturing processes, etc.

Since the parameters defining the actual load or load-carrying capability in design will not correspond to those actually experienced, the variables L_a and L_c should be defined by their mean values, $\bar{L}_a$ or $\bar{L}_c$, and a tolerance band on the mean values:

$$L_a = \bar{L}_a \pm \Delta L_a \tag{5.1}$$

$$L_c = \bar{L}_c \pm \Delta L_c. \tag{5.2}$$

These variables are not deterministic but random in character. The mean values and standard deviations are dependent on the distribution curve corresponding to the frequency of occurrence. These variables are independent. The objective is to establish a relationship between $\bar{L}_a$ and $\bar{L}_c$ such that their ratio is defined as the factor of safety:

$$S = \frac{\bar{L}_c}{\bar{L}_a} \tag{5.3}$$

Now, if the life of a system is assumed to be a random variable with a distribution function $\phi(t)$, which is the probability that the system or element fails before time t_1, the average life could be defined as the mean of the distribution $\phi(t)$, and the probability of operation without failure within the interval (t_0, t_1) with $t_1 > t_0$:

$$\frac{\phi(t_1) - \phi(t_0)}{1 - \phi(t_0)} . \tag{5.4}$$

The system or element is assumed to be in good operating conditions at time t_0.

Consider now the probability of failure-free operation within the interval $(t, t + \epsilon)$ with ϵ being infinitesimal. If $f(t)$ is the probability density function corresponding to $\phi(t)$, then 'the hazard rate', also called 'instantaneous failure rate', is given by

$$F_i(t) = \lim_{\epsilon \to 0} \frac{\phi(t + \epsilon) - \phi(t)}{\epsilon[1 - \phi(t)]} = \frac{f(t)}{1 - \phi(t)} . \tag{5.5}$$

The probability of failure-free operation until time t, or the probability of surviving life, called *reliability*, is given by

$$R(t) = 1 - \phi(t). \tag{5.6}$$

It follows from equation (5.5) that

$$f(t) = \frac{d[\phi(t)]}{dt} \tag{5.7}$$

and

$$F_i(t) = \frac{d}{dt}(\log[1 - \phi(t)]). \tag{5.8}$$

A simple integration gives

$$\phi(t) = 1 - \exp\left[-\int_0^t F_i(w)dw\right] \tag{5.9}$$

and

$$f(t) = F_i(t) \exp\left[-\int_0^t F_i(w)dw\right]. \tag{5.10}$$

5.2. Structural integrity

Any manufacturing process is known to introduce defects and internal stresses in structural elements. There are also inherent flaws within the materials. The discipline of fracture mechanics postulates that every structure or material contains initial crack-like imperfections. These cracks are not necessarily harmful until they grow to the point where the structural integrity is threatened. Therefore, it is essential to design for the remaining life of damaged structural components [1].

In order to have a knowledge of the degree of criticality of components, it is necessary to consider the follow:

(1) an effective inspection procedure after manufacturing and during in-service operations;
(2) the type of materials used in manufacturing and size of each part;
(3) the static and dynamic safety factors adopted in the design;
(4) the stress concentrations; and
(5) the load transfer characteristics.

The data supplied by testing and from in-service operations are of ultimate importance for a correct estimate of the degrees of criticality. Quality assurance during manufacturing is also crucial such that the distribution and localization of preexisting cracks can affect the life of components. This information could be obtained through a statistical analysis of tests of the prototype or tests of other similar structures. The stress spectrum derived from the load spectrum provides the limits for the in-service operation and the requirements for periodic inspection. To this end, the linear elastic fracture mechanics (LEFM) theory based on a critical stress intensity factor can be used:

$$K_I \geqslant K_{Ic}. \tag{5.11}$$

For complex and large size structures where failure could lead to serious loss of human life and/or costly properties such as airplanes, ships, dams, etc., it is highly advisable to install instruments at locations where cracks are likely to propagate such that corrective measures can be made prior to catastrophic failure. Since the quality and reliability of a system is determined during the design phase, the manufacturers and users of aircrafts have been concerned with the structural integrity. Owing to newly established ideas and approaches, some of the past methods were bypassed. The 'safe life' concept consisting of replacing parts after a period of operation is no longer adopted. This decouples the requirements of safety and durability. The new policy adopted by the United States Air Force (USAF)[2] and defined in the specification MIL-STD-1530A is a good example. This new approach states that aircrafts shall be designed:

(1) from the 'safety' viewpoint
    (a) with tolerable damage associated with fail safe, or
    (b) with safe crack growth; and
(2) from the 'durability' viewpoint with an economic life greater than the design life as shown in Figure 5.1.

The economic life (EL) is characterized by a high derivative of the increasing rate of cracking in the tested element which can be considered directly proportional to the repair cost. EL should be greater than the design service life.

It is not easy to define durability as it is related to quantifying the level of acceptance

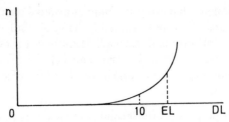

Figure 5.1.   Durability concept. n = number of cracks (repair cost); DL = design lifetimes;
EL = economic life.

of defects and to demonstrate that the design fulfills this objective. The fatigue strength and the damage tolerance could be calculated from a probabilistic analysis but the durability limits are not so easily determined. A first indication can be given by the cyclic test when widespread crack propagation takes place.

5.3. Designing for structural integrity

If the designer lacks knowledge in material behaviour and working conditions of the component, obviously no reliable information in design could be found even if the analysis were correct. Needless to say, input data are the prerequisites for obtaining a reliable solution.

Some of the important considerations are:

(1) Based on the correct working conditions, stress calculations can be made to predict the fatigue strength of structures by analytical or numerical means such as the finite element method.
(2) A scale model could be made for performing experimental stress analysis. This serves as a check of the analytical prediction in (1).
(3) The experimental stress analysis approach can be used to reproduce the working conditions and to yield useful information on fatigue life by improving on the strength, stress concentrations, residual stresses, etc.
(4) The construction of a prototype can be done after the corrections and improvements made in accordance with the results given by the experimental stress analysis. At this stage, the designer can obtain more accurate information on the real loads and fatigue life.
(5) After the first prototype test, other tests can follow with different experimental variables in order to get information about the durability. At this stage, the design could be finalized.

The above five (5) stages may appear to be time-consuming but they can lead to considerable saving in time and cost.

5.4. Safety factor and reliability

Recalling the definition of the safety factor as indicated in equation (5.3), alternatively, s can also be defined as

172

$$s = \frac{\bar{\sigma}_{rup}}{\sigma_{ad}} \tag{5.12}$$

in which $\bar{\sigma}_{rup}$ is the strength of the element taking into consideration the manufacturing process and other factors that influence its behavior. It could be the ultimate strength, the critical stress for buckling, etc. $\bar{\sigma}_{ad}$ is the allowable or working stress. Note that the meaning of s is only meaningful when defined in terms of the mean value and a standard deviation.

Once the distribution functions of the random variables in design are known, it is easy to find s. To this end, it is necessary to know:

(1) data on material behavior used in manufacturing;
(2) data on component behavior within the system;
(3) data on working conditions such as the load spectrum; and
(4) data on distribution and location of defects.

The above data shall be subjected to appropriate statistical analysis. This includes the conversion of data to mathematical equation forms and the adequate fitting to the distribution curves. They will serve as the inputs to the construction of codes. Such codes will, of course, be different from those established from the deterministic analysis. The probabilistic approach to failure could be defined in different ways involving the simple subsystem failure and the global failure. Some of the details will be examined.

Specimen tests

It is essential that test data are representative of the physics of the problem. The specimens should be reproduced from the actual equipment and follow the specifications required in manufacturing. Reproducibility of data should be kept in mind. This requires a minimum variation in test conditions including machining and controlling test conditions so that scattering in data can be minimized. The number of test specimens required to establish a statistically valid sample is subject to controversy. To test as many specimens as possible is desirable in order to ensure maximum confidence; but this is usually limited by economic considerations, especially in the case of complex specimens. There is no suitable answer to this problem and no mathematical technique is available for determining the sample size [3]. Smaller samples are preferred. As a rule, 20 to 30 specimens are sufficient to determine the average values of data according to standard deviations. The number of test specimens will depend therefore on the rate between the cost of rejection of the sample and the cost of inspection.

Confidence levels

The purpose of the analysis is to calculate the design strength of a material having a known reliability and a known degree of confidence. Assume that the distribution curves of strength for a certain part for two groups of test samples, A and B, are Gaussian. Due to variations in materials, manufacturing, test techniques, etc., the sample distribution curves A and B do not coincide with the universal distribution curve. If a certain value of reliability is selected, the percentage of the tests will exceed

173

this value. However, in some samples, only a certain percentage of the tests will exceed the prescribed level. The percentage of samples in which the reliability level of tests exceeds the prescribed level is called the *confidence level*. For example, the requirement of 99% reliability at 95% confidence level means that, if a large number of samples each consisting of 100 test specimens is tested, 99 specimens in each sample would exceed the design level for 95% of the samples; only in 5% of the samples there would be less than 99 specimens exceeding the design level.

The distribution curves have different shapes corresponding to the random nature of data compiled from the tests. The assumption that strength data fits a normal distribution is usually accepted. This distribution is completely defined by the mean and the standard deviation. Other distributions such as Weibull, lognormal, exponential have been used as they may fit the test data better. Test data is never normally distributed. Such an hypothesis can be statistically tested either qualitatively or quantitatively [4]. The qualitative test is made by visual inspection of hystograms. The χ^2 test is a quantitative method of testing normality of data.

It is especially important not to take into consideration bad data, since these data influence the results, particularly the standard deviation. The rejection of such data demands a careful study of abnormal values, that usually indicate imperfection in the material or poor tests. If no flaw is found in the material and the test appears to be normal, then the Chauvenet's criterion [5] could be applied to reject a test value.

Allowable stress

The calculation of the design-allowable stress requires the estimation of lower tolerance limit based on the mean value and the number of standard deviations between mean and this limit for various confidence levels and reliabilities.

References

[1] White, D.J. and Gray, T.D., 'Damage Tolerance Assessment of the A.7D Aircraft Structure', 5th International Conference on Fracture, Cannes (1981).

[2] Wood, H.A., 'Structural Integrity Technology for Aerospace Applications', Conference on Structural Integrity Technology, ASME, New York (1979).

[3] Smith, F.C., 'The True Design Strength of Materials and Joints', Machine Design (December 8, 1966).

[4] Sinha, S.K. and Kale, B.K., 'Life Testing and Reliability Estimation', Wiley Eastern Limited, New Delhi (1980).

[5] Vaugh, A.E., 'Elements of Statistical Methods', McGraw-Hill, New York (1952).

SUBJECT INDEX

Accelerated life, 142
Aluminum, 5, 7, 10, 79
    fatigue data, 16, 17, 76
    material properties, 5, 6, 7, 42, 77
    specimens, 6, 24, 25
Anisotropy, 91
ASTM, 36, 42, 159
    specimens, 89
    thickness requirement, 90, 94

Brittle fracture
    limitation, 95, 159
    theory, 36

Central crack
    composite, 21
    idealized specimen, 45, 46, 47
    metal alloy, 24
Compliance, 43
Composite (*see* graphite-epoxy)
Concentrated loads, 63
Core region, 41
Corrosion, 92, 113, 124, 134, 159
Crack front
    curvature, 63, 67, 89
    slanted, 59
Crack initiation (*see* direction of crack growth)
Crack growth rate, 3, 5, 72, 74, 75, 137, 156
    aluminum, 16, 17, 76
    steel, 15, 75, 135, 136
    titanium, 14, 77
Crack nucleation (*see* damage model)
Crack opening displacement, 94, 147, 159, 167
    criterion, 98, 99
    measurement, 90
Crack retardation, 10, 86, 158
Crack tip (*see* stress field)
Curved crack (*see* crack front curvature)
Cyclic loading, 2, 3
    constant amplitude, 3, 8, 10, 11, 71, 72, 133, 135, 140, 163

    random, 112, 125, 129
    spectrum, 9, 10, 11, 131, 140
Cylindrical bar, 49

Damage model
    accumulation, 71, 72, 110, 132
    statistical (*see* statistical analysis)
    tolerance assessment, 155, 167, 172
Defect size
    description, 152
    three-dimensional, 160
Dilatational energy density, 37, 39
Direction of crack growth (*see also* mixed mode)
    initiation, 41, 46, 58, 59, 60
    propagation, 7, 83
Distortion energy density, 37, 39
Ductile behavior
    crack front curvature, 63, 67, 69
    fatigue, 70
    fracture criteria, 94
    size effect, 68
Dynamic loading, 124, 125, 126

Edge effect, 44
Elastic-plastic effects, 40, 65, 94, 107
Energy release rate, 36, 43

Fatigue
    analytical models, 3, 5, 72, 73, 74, 75, 78, 137, 163
    data (*see* crack growth rate)
Fiber volume fraction, 8
Flaw acceptance criteria, 151, 155, 158, 162, 167
Flaw size (*see* defect size)
Fracture failure criteria
    crack opening displacement, 98, 99, 159, 167
    Griffith and Orowan, 95
    J-integral, 96, 97, 98, 159, 167

Strain energy density, 3, 5, 40, 71, 167
Fracture surface
  angle crack in fatigue, 18, 19
  disc brake, 115
  rail head, 146, 147
  wheel, 104, 105, 106
Fracture toughness test
  loading, 90
  specimens, 89
Fracture toughness values (*see* aluminum,
    steel and titanium)
Fretting, 134

Gaussian process, 109, 117, 118, 119, 120,
    122, 123, 126, 128
Graphite-Epoxy, 11, 12
  analytical model, 8, 21
  fatigue data, 22, 23, 24, 29, 30, 32, 33
  material properties, 7
Griffith surface energy, 95

Heated pipe, 55
Hysteresis energy density, 70, 71

Impact load, 123
Incremental growth, 40, 63, 65
Inhomogeneity, 91
Initial defect, 48, 50, 152, 157, 166
In-plane extension, 45
In-plane shear, 46
Interlaminar matrix crack, 8, 10

J-integral (*see* fracture failure criteria)

Ligament
  length, 41, 63, 87, 99
  size, 40, 69
Linear damage
  summation, 132, 133, 140
  strength reduction, 10, 11, 173
Load influence
  factor, 129
  sequence, 2, 107
  spectrum (*see* cyclic loading)
Location of failure
  fracture, 40
  yielding, 40, 41

Maneuver load, 127, 128
Mean parameter
  life, 11
  stress, 3, 9, 76, 78, 109
  stress intensity factor, 76
Mixed mode
  cylinder in torsion, 49
  fatigue, 3, 4, 17, 20, 81, 84, 85

pressurized vessel, 57
shell in torsion, 59, 60
slanted crack, 61, 62

Nondestructive testing (NDT), 151, 153,
    154, 156
Nonlinear behavior, 68

Out-of-plane shear, 47
Overloads, 11, 12, 86, 110

Path dependent process, 69, 70
Penny-shaped crack, 48
Plane strain, 39, 90, 94
Plastic deformation (*see* elastic-plastic
    effects)
Plastic length, 95
Pop-in, 90
Press-fit, 51
Pressurized vessel, 57
Probability density function, 110, 152, 153,
    170
Proof test, 142

Raleigh process, 110
Random load (*see* cyclic loading)
Reliability, 170
Resistance curve, 98
Root-mean-square (RMS), 109, 110, 113,
    117, 118
Rotating disc, 53

Safety-factor, 173
Self-similarity, 98
Shakedown, 107
Shape change, 37, 39
Singular behavior, 40, 86, 87
Size effect, 36, 62, 66, 67
Spectral analysis, 111, 113
Spectrum load (*see* cyclic loading)
Stable growth, 40, 63, 65, 67
Statistical analysis
  probabilistic, 11, 170
  root-mean-square, 109
Steel
  fatigue data, 26, 75, 135, 136
  material properties, 5, 6, 42, 78, 88
  specimens, 6, 24, 25
Stiffened parcel, 56
Strain energy density criterion, 40, 71
Strain energy density factor, 87
  angle crack, 57
  central crack, 46, 47
  critical values, 42, 45, 66, 69
  penny-shaped crack, 49
  range, 3, 4, 14, 15, 16, 17, 71, 72, 74, 75,
    76

slanted crack, 59, 60
Strain energy density function, 37
    critical values, 38, 40, 42
    dilatational component, 37, 39
    distortional component, 37, 39
    stationary values, 40, 41
Strength test, 38
Stress field, 86, 87
Stress intensity factor
    angle crack, 7, 57, 80
    central crack, 43, 45, 46, 47
    critical values, 42, 49, 66, 88
    penny-shaped crack, 48
    range, 73, 76, 137, 163
    slanted crack, 59, 60
Surface layer energy, 7
Surface flaw, 25, 27, 28, 160

Temperature effect, 91
Testing
    accelerated life, 142
    proof, 142

Thermal stress, 92
Titanium
    fatigue data, 14, 77
    material properties, 42, 78
    specimens, 84, 85
Tolerance assessment (*see* damage model)
Torsion, 38, 49, 58, 59

Ultimate strength, 5, 7, 11, 48, 52, 173
Unstable fracture, 40, 63, 69, 90

Volume damage, 37, 39

Weldment
    joint, 160, 165
    fatigue data, 164, 165
    structure, 108
Wheel fracture (*see* fracture surface)

Yield strength
    influence, 88, 90, 94, 95
    material property, 5, 42